# SECRECY AND SCIENCE

# Secrecy and Science

## A Historical Sociology of Biological and Chemical Warfare

BRIAN BALMER
*University College London, UK*

LONDON AND NEW YORK

First published 2012 by Ashgate Publishing

Published 2016 by Routledge
2 Park Square, Milton Park, Abingdon, Oxfordshire OX14 4RN
711 Third Avenue, New York, NY 10017, USA

First issued in paperback 2016

*Routledge is an imprint of the Taylor & Francis Group, an informa business*



**British Library Cataloguing in Publication Data**
Balmer, Brian, 1965–
Secrecy and science : a historical sociology of biological and chemical warfare.
1. Defence information, Classified—Great Britain—History—20th century. 2. Biological weapons—Research—Government policy—Great Britain—History—20th century.
3. Chemical weapons—Research—Government policy—Great Britain—History—20th century. 4. Freedom of information—Great Britain—History—20th century.
I. Title
355.3'433'0941'0904–dc23

**Library of Congress Cataloging-in-Publication Data**
Balmer, Brian, 1965–
Secrecy and science : a historical sociology of biological and chemical warfare / by Brian Balmer.
p. cm.
Includes bibliographical references and index.
ISBN 978-1-4094-3056-8 (hbk.:alk. paper)
1. Biological warfare—Research—Great Britain—History. 2. Chemical warfare— Research—Great Britain—History. 3. Official secrets—Great Britain—History. 4. Research, Military—Social aspects. I. Title. II. Title: Historical sociology of biological and chemical warfare.
UG447.8.B325 2012
358'.34072041–dc23

2011039534

ISBN 13: 978-1-138-27728-1 (pbk)
ISBN 13: 978-1-4094-3056-8 (hbk)

# Contents

# List of Tables

# List of Abbreviations

| | |
|---|---|
| BBC | British Broadcasting Corporation |
| BRAB | Biological Research Advisory Board |
| BW | Biological Warfare (sometimes Biological Weapon) |
| CBW | Chemical and Biological Warfare (sometimes Chemical and Biological Weapons) |
| CDEE | Chemical Defence Experimental Establishment |
| CND | Campaign for Nuclear Disarmament |
| DRPC | Defence Research Policy Committee |
| CW | Chemical Warfare (sometimes Chemical Weapon) |
| HMG | His/Her Majesty's Government |
| HSP | Harvard-Sussex Program |
| ITV | Independent Television |
| IRA | Irish Republican Army |
| MoD | Ministry of Defence |
| MP | Member of Parliament |
| MGO | Master General of the Ordnance |
| MRC | Medical Research Council |
| MRD | Microbiology Research Department |
| MRE | Microbiology Research Establishment |
| R&D | Research and Development |
| SIPRI | Stockholm International Peace Research Institute |
| TNA | The National Archives (Kew, UK) |
| UK | United Kingdom |
| US | United States |
| USA | United States of America |

# Preface and Acknowledgements

This book is the second part of an unintended trilogy. My previous book, *Britain and Biological Warfare*, charted the history of the UK biological warfare programme from its inception to the mid-1960s. My intention after finishing the book had been to mop up and then move on to a parallel historical account of the UK chemical warfare programme. The mopping up exercise consisted of accessing some more recently declassified and released archival material with the intention of turning it into a series of articles. Dealing with this new material was somewhat dissimilar to my previous research, which had been aided by the chronological organisation of sources in the UK National Archives, and so made it relatively easy to tell an unbroken historical narrative. In contrast, the left-over archival sources were, by-and-large, single government files containing several documents, or short series of government files, each concerning particular incidents such as accidents, scares or disputes over policy. Not surprisingly, many of these documents were originally classified as Top Secret or Secret. In the course of researching and thinking about this archival material, I kept returning to the theme of secrecy and what academic social scientists might have to say about it. What is the nature of secrets? How does secrecy operate as a social process? What is the relationship between secret knowledge, open knowledge and epistemology? The result of my mopping up and subsequent foray into these questions about secrecy is this book.

*Secrecy and Science* is, therefore, a series of historical case studies, which largely focus on the UK biological weapons programme in the Cold War. The UK's chemical warfare programme also features, particularly in the later chapters. As such, the book can be read straightforwardly as an account of certain events in this history and I know that some readers will just want to engage with that narrative. For them, I hope to have provided sufficient clarity in the case studies to side-step the discussions about secrets, although I hope they will not ignore them entirely. This is because there is a broader purpose to this book. Historical and contemporary case studies do not, I believe, generate rigid laws of social theory and this is most certainly not my aim in this book. Case studies do, however, provide what sociologist Jennifer Platt has called a 'barium meal' through social processes, highlighting and sensitising us to features we might recognise as being present or conspicuously absent in analogous cases (Platt 1988). It is this more modest form of claim about secrecy that I will make and that should interest a second group of readers, who are primarily my colleagues in the field of Science and Technology Studies, which, using the University College London definition, encompasses history, philosophy and social studies of science. What I have to say about secrets and science should also be of interest for other social scientists with an interest in secrecy.

In the chapters that follow, my background in Science and Technology Studies, and the sociology of science in particular, has strongly informed my analysis of the case studies. Put rather simply and briefly, this approach regards knowledge as a social institution. There is a process involved in coming to claim to know something, and for the sociologist of science this is a social activity, as much as it is a cognitive or technical or materially grounded activity. A few critics have sometimes read the central ideas in Science and Technology Studies as a set of 'science is nothing but' claims, that we don't know anything, or that all knowledge is the same, or even that it is a mirage. While the Science and Technology Studies approach has quite fundamental epistemological implications, claiming that knowledge is in some sense 'constructed' is really not equivalent to any of these 'nothing but' claims. For readers unfamiliar with social studies of science, there are several excellent introductions to this field which provide a clearer, lengthier and more robust defence of this approach than I have space for here (Bucchi 2004, Erikson 2005, Sismondo 2010, Yearley 2005). For those colleagues within Science and Technology Studies who read this book, it will raise issues around the fragmentation of knowledge which I believe have been largely neglected in our collective attempt to make sense of what it means to claim to know something. Certainly, I will be arguing that the question 'who knows what, where?', a question at the heart of the relationship between knowledge and secrecy, is central to Science and Technology Studies and so worth pausing over and spending time to think about.

An earlier version of chapter two appeared as 'Killing "Without the Distressing Preliminaries": Scientists' Defence of the British Biological Warfare Programme', *Minerva* 4(1), 2002, 57–75. I am grateful to Springer for their permission to reproduce material from that article in this book.

An earlier version of Chapter 3 appeared as 'How Does an Accident Become and Experiment? Secret Science and the Exposure of the Public to Biological Warfare Agents', *Science as Culture* 13(2), 2004, 197–228. I am grateful to Taylor and Francis for permission to reproduce material from that article in this book.

An earlier version of Chapter 4 appeared as part of 'How Does Secrecy Work? Keeping and Disclosing Secrets in the History of the UK Biological Warfare Programme', in *A Web of Prevention: Biological Weapons, Life Sciences and the Governance of Research* edited by B. Rappert and C. McLeish. 2007, 173–188. I am grateful to Earthscan for permission to reproduce material from that chapter in this book.

An earlier version of Chapter 7 appeared as 'A Secret Formula, A Rogue Patent and Public Knowledge about Nerve Gas: Secrecy as a Spatial-Epistemic Tool', *Social Studies of Science*, 35(5), 2006, 691–722. I am grateful to Sage whose re-use policy generously allows me reproduce material from that article in this book.

Several people deserve thanks for their help, both direct and indirect, in bringing this book together. An enormous thank-you must go to Julian Perry-Robinson, who introduced me to the whole field of chemical and biological weapons control, for his continuing encouragement and willingness to share his tremendous expertise.

Brian Rappert deserves special mention for our many discussions about biological warfare and secrecy. Thank you also to Gail Davies for helping me to think like a geographer, at least a little bit. Hugh Gusterson, Jeanne Guillemin and Sheila Jasanoff were superb hosts during a month-long visit to MIT in 2005, where some of the chapters of this book began to take shape. The final stages of the book were completed while I was a visiting researcher in the Departamento de Historia y Filosofía de la Ciencia, la Educación y el Lenguaje, at the University of La Laguna, Tenerife, with support from the Spanish Ministry of Science and Innovation Research Project FFI2009-09483. My thanks to Amparo Gómez and Antonio F. Canales for hosting this visit. The chapter on ignorance and secrecy only saw the light of day thanks to the encouragement of Linsey McGoey. Then I have a list of people to whom I'm indebted for helpful discussions about various aspects of this book: Donald Avery, Andrew Barry, Robert Bud, Joe Cain, Gradon Carter, Hasok Chang, Malcolm Dando, Daniel Feakes, Donald Gilles, Mike Goodman, Catherine Jefferson, Inga Kroener, Greg Koblenz, Filippa Lentzos, Jez Littlewood, Caitriona McLeish, Roy McLeod, John Moon, Norma Morris, Paul Nightingale, Graham Pearson, Ulf Schmidt, Nicholas Sims, John Stone, Stephen Twigge, John Walker, Mark Wheelis, Neal White, Henrietta Wilson, and Sally Wyatt.

I am lucky to work in an intellectually inspiring environment and I owe a particular debt of gratitude to my colleagues and students, past and present, in the Department of Science and Technology Studies at University College London. Consider yourselves all thanked here. But I do want particularly to mention the colleagues who taught with me on our undergraduate course 'Science, Warfare and Peace' over the past fifteen years – Jon Agar, Arthur Miller, Charles Thorpe – I have learnt much from you that has found its way into this book. And likewise, thanks to all my PhD students, past and present, who have also taught me so much. Beyond the department, Claire Jarvis and Bethan Dixon at Ashgate have been helpful and patient editors. Gary Woodward and Neil Roscoe, thanks for all sorts of help, friendship and support – I hope you enjoy the book. Eileen and Leslie Balmer get my thanks for being my wonderful parents. And Kevin Hannigan gets my thanks too, for everything.

Brian Balmer
August 2011

# Chapter 1
# Secret Science

Secrets are rarely straightforward acts of concealment and revelation. Listen to two anecdotes about secrecy that I was told recently by someone who worked for the UK aerospace industry in the early 1960s. "My first job", he recalled, "was as a junior technical officer working in a team trying to understand the aerodynamics of ship-to-air missiles. We would wire up the missiles to analogue computers and simulate their in-flight behaviour. The work was highly compartmentalised, and I was given no information about many of the other components of the missile. All the work was classified on a scale ranging through unrestricted, restricted, confidential, secret, secret discrete and top secret; as a junior technical officer, I was cleared to look at material up to the level of secret. I remember what happened with one report that I'd written with several of my colleagues that was classified secret, and then sent off to the company library. A few months later, our own copy of the report wasn't to hand, so one of our team volunteered to fetch the copy from the library. Now, despite being one of the authors, with full knowledge of the contents of the report, he was refused permission to borrow the report on the grounds that he was a senior technical assistant and therefore not cleared to the requisite level of access."

"Later, still working as a junior technical officer for the same company," he continued, "I was moved from simulation onto a contract to refurbish US ground-to-air missiles. We had a delay of around a month before the missiles actually arrived, even though we had been sent the manuals. Someone even put up a sign in our corner of the laboratory saying 'Men Without Missiles', as we sat around not able to get started. All our attempts to get the Americans to send more information about the missiles met with refusal. Anyway, during this wait, one of my colleagues turned up to work with a hobbyists' scale model of the missile and launcher; this was something that was widely available to buy in toy shops. Not that long afterwards, when the real missiles arrived for refurbishment, we realised that the hobby model was incredibly accurate, even down to the number of rivets on the missile wings." Someone on the inside inadvertently finds himself on the outside; people excluded from a secret later find they had unknowingly accessed the inside by the most unlikely means. Secrecy, it seems, does not simply separate the world into a neat inside and outside.

This book picks up the themes of secrecy and science and traces some of the contours of their relationship through more focussed empirical material. The historical focus is scientific research and government policy on biological and, to a lesser extent, chemical warfare during the Cold War. The geographical focus is the United Kingdom. My aim is to use case studies to recount some intriguing aspects of Cold War chemical and biological warfare policy, but also to make broader observations about secret science. I have explained the choices that led to the book and the relationship between these broad and narrow spotlights on a more personal note in the preface. That explanation is particularly relevant to any readers who are unfamiliar with the ideas and approaches emerging from the academic field of Science and Technology Studies.

Secrecy has never been a major topic of research in social science, although it has not been entirely neglected. Recent bibliometric data supports this claim (Marx and Muschert 2008).[1] With respect to secrecy and science, much of the scholarly discussion focuses on the ethics of secrecy rather than on a sociology of secrecy that considers its enactment, meanings and effects. Academic sociology and history of science frequently portrays secret science as similar to open science, just done behind closed doors. It is such oversimplification that will be challenged throughout this book. In order to do so, this opening chapter will provide background contextual information: a brief overview of chemical and biological warfare and its history; a consideration of some of the social scientific writing on secrecy and science; and a brief consideration of sources and historiographical issues in researching secrets and in researching the history of chemical and biological warfare. My introduction will finish with an outline of the book and main arguments from each chapter.

## A Brief History of UK Chemical and Biological Warfare Research and Policy

The case studies in this book focus on the British biological and, to a lesser extent, chemical weapons programmes, which were some of the largest in scale and scope in the twentieth century. Although the chapters are ordered with a sense of chronological progression, I do not intend to provide a detailed time-line. Instead, in order to give you some degree of orientation, this section provides a brief overview of key developments in research and policy in this field in the UK.

Biological weapons are conventionally defined as those which employ living organisms – usually bacteria, viruses or fungi – to cause disease in humans, other animals and crops. Disease-causing micro-organisms, also known as pathogens, responsible for anthrax, plague and tularemia are just three examples of the many diseases that scientists have researched as potential biological weapons in the past (Jones 2010, Wheelis et al. 2006). Chemical weapons generally refer to inert chemical agents whose primary mode of causing harm is their toxicity, thus

1 See Marx and Muschert (2008) footnote 1 for bibliometric evidence on the relative paucity of social scientific literature on secrecy. This neglect is exemplified in their observation that secrecy and secret societies have no entries in the *Penguin Dictionary of Sociology* (1994), *Concise Oxford Dictionary of Sociology* (1994), *Cambridge Dictionary of Sociology* (2006), *Encyclopaedia of Sociology* (2000), *Encyclopaedia of Social Theory* (2005), or *Blackwell Companion to Major Social Theorists* (2000).

marking them out from other weapons such as incendiary devices. As such they may blister, choke, poison the blood, and interfere with the nervous system to kill or incapacitate (Kenyon 2000). Agents such as chlorine and mustard gas, used in the First World War, were accompanied by later generations of highly toxic and fast-acting nerve gases that target the nervous system. Although poison weapons – both biological and chemical – have a long history dating back to ancient times, the rise of the chemical industry in the nineteenth century allowed for mass production of chemicals for use in war; also in the late nineteenth century the advent of the 'germ theory' of disease, which postulated a single organism as the causative agent of any disease, underpinned thinking about biological warfare (Coleman 2005, Mayor 2003, Perry-Robinson 1981, Perry-Robinson and Leitenberg 1971) .

Various authors have characterised the defining features of twentieth century warfare as the impersonalisation and industrialisation of mass killing, together with the growth of markedly closer links between science and the military (Edgerton 1990, Glover 1999, Mendelsohn 1997, Roland 2003). Certainly, the first large-scale lethal chemical warfare attacks during the First World War fit with this characterisation. Germany had tried initially, and with very limited effect, to make use of irritant gases in the war, but from the end of 1914 moved on to search for lethal agents. Scientist Fritz Haber directed this work. Drawing on the industrial might of the German chemical industry, he was able to make use of poisonous chlorine, which was routinely produced in the German dyestuffs industry. Haber was interested not only in the details of the science; he was also involved at the Front in the organisation of the first use of chlorine as a mass casualty weapon (Charles 2005). In April 1915 the Germans launched a series of attacks using cylinders of pressurised liquid chlorine, released as a huge yellow-green toxic cloud, on the Western front at Ypres in Flanders, Belgium. By September, the British were in a position to retaliate in kind, launching an attack at Loos in Belgium (Harris and Paxman 1982, Haber 1986, Carter 2000).

In the following months, Porton Down in Wiltshire was selected to be the home of the British chemical warfare research programme. The earliest activities on the site, named the War Department Experimental Station (and briefly beforehand as the Experimental Ground), took place in 1916. Further north, an old factory at St Helens in Lancashire was acquired as a process research plant to study large-scale production of agents. This site eventually became known as the Chemical Defence Research Establishment Sutton Oak. It had production facilities, but its main role was to develop new large-scale production processes, the details of which could then be transferred to the chemical industry. Throughout the rest of the war, the UK response to the new chemical warfare engaged a far wider network of organisations, both civil and military. Porton's official historian, Gradon Carter, points out that the network included: 'the chemical industry, the Ministry of Munitions, the Royal Society and the Services. Defensive matters were largely the concern of the Anti-Gas Department in London' (Carter 2000:2).

The inter-war years saw new attempts to ban chemical weapons. It is not wholly mistaken to say that these were provoked by revulsion to these weapons during the war. But the idea of widespread revulsion glosses over the history somewhat as 'in 1919 … those of the general public who could recall anything of the wartime publications on CW might have adopted any one of a number of assessments: gas as a humane weapon, gas as a terror weapon, gas as just another weapon as horrible as any other' (SIPRI Vol 1: 234). The build-up of support for a ban is best told about elsewhere (McElroy 1991, SIPRI 1971, Spiers 2006). It culminated, through negotiations by the League of Nations, in the 1925 Geneva Protocol, a treaty which placed a ban on both chemical ('use in war of asphyxiating, poisonous or other gases, and of all analogous liquids, materials or devices') and, as something of an afterthought, 'bacteriological' warfare. The treaty was far from ideal. What counted as 'use in war' remained unspecified. Significantly, some nations, notably the USA, did not ratify the treaty, while many others tabled reservations. In consequence, the Protocol was effectively an agreement to 'no-first-use' of these weapons rather than an outright ban.

The UK Cabinet maintained its commitment to the Geneva Protocol in the build-up to the Second World War, but permitted offensive research and development; it also approved production and stockpiling of mustard gas (Carter 2000). At the re-named Chemical Defence Experimental Station, Porton saw an influx of scientists from industry and academia once war broke out. Here, research focused on new agent development and field testing. Beyond Porton, factories run by Imperial Chemical Industries (ICI) produced phosgene, mustard gas and a tear gas (bromobenzyl cyanide) as weapons (Carter 2000).

Concern about the possibility of biological warfare had grown in public and behind the closed doors of government during the 1930s, but in the UK this had led to little beyond the stockpiling of vaccines and sera as precautionary measures. The war engendered fresh consideration of the threat, particularly by senior civil servant Maurice Hankey, and resulted in the establishment of the Biology Department Porton in October 1940. Hankey, acting for Churchill, instructed scientists at the Biology Department, as a matter of urgency and utmost secrecy, to create a biological weapon that could be deployed at short notice in retaliation against a similar attack (Balmer 2001). Throughout the war, the team at Porton worked on various aspects of this experimental form of warfare and eventually produced two potential weapons: an experimental anti-personnel anthrax bomb, and some five million cattle-feed cakes containing anthrax spores, ready, but in the event never used against livestock (Carter and Pearson 1999).[2]

For a brief period after the Second World War, chemical, biological and atomic weapons research were each accorded equally high priority in UK defence policy. Atlee's Labour government resolved in 1946 to be 'in a position to wage chemical

2 Chemical weapons were not used on the battle field in the Second World War. Some use was made of biological weapons by the Japanese, with very limited success in developing a bomb, the head of the programme, Ishii, resorted to tactics such as scattering civilian areas with plague-infected fleas (See Harris 1994, Guillemin 2005, 84–86).

warfare from the start of hostilities'.[3] This decision was taken ahead of the Cabinet decision in January 1947 that Britain would proceed with building its own atomic bomb. Later that year, the Chiefs of Staff, referring explicitly to atomic, biological and chemical weapons, agreed on 'a cardinal principle of policy to be prepared to use weapons of mass destruction. The knowledge of this preparedness is the best deterrent to war in peace-time'.[4]

The post-war continuation of the high priority accorded to chemical weapons research was spurred by the Allied discovery of the German nerve gases, which acted by disrupting the functioning of the nervous system. These new gases had been discovered in the course of pesticide research in Germany just a few years prior to the outbreak of war, but had remained unused. They were found by Allied troops in the closing stages of the war when they came across German munitions filled with agent Tabun. The new agents, such as Tabun and Soman, collectively named the G-agents, were far more aggressive, combining a much greater toxicity and speed of action, than anything possessed by the UK. Work to assess these novel agents became the major focus of the Chemical Defence Experimental Station – renamed the Chemical Defence Experimental Establishment (CDEE) in 1948. CDEE was also responsible for the development of a new series of highly toxic nerve agents, the V-agents, discovered in the early 1950s (see chapter seven, McLeish and Balmer 2012). Work also continued at Sutton Oak until 1952 when the facility was transferred to the more isolated site at Nancekuke in Cornwall.[5]

The UK biological warfare research programme also received generous support and funding. The scope of the research at the re-named Microbiological Research Department was broadened to take in a wider range of pathogenic organisms, with work ranging from fundamental studies to open air trials of pathogenic organisms at sea. In terms of policy, the Chiefs of Staff noted that 'for the successful development of defensive measures it was essential to study the offensive field'.[6] In this respect, an Air Staff Requirement was put in place for Porton to develop an anti-personnel biological bomb (Balmer 2001). Throughout this period, scientists in the British research programme for both biological and chemical warfare maintained extremely close links with their counterparts in the US and Canadian programmes, an arrangement that had been established during the war and became known as the Tripartite Agreement (Carter and Pearson 1996).

The status of chemical and biological weapons in UK policy and planning did not remain a high priority for long. Several factors contributed to this change, not least the first Soviet atomic bomb explosion in 1949, followed by Britain's

3 TNA [Kew, The National Archives], CAB[inet office papers] 131/1. Minutes of Defence Committee of the Cabinet, 20 June 1946.

4 TNA, DEFE[nce ministry papers] 10/19. DRPC. Final Version of Paper on Future of Defence Research Policy, 30 July 1947.

5 Nancekuke was closed in 1979.

6 TNA, DEFE 10/26. DRP(50)53. DRPC. BW Policy. Note by the Chairman, BW Sub-Committee (11 May 1950).

first successful test of their own atomic bomb in 1952. With British nuclear weapons now a reality, the decision by the Cabinet two years later to pursue a hydrogen bomb contributed further to the weakening of interest in biological or chemical weapons. If these nuclear developments did not seal the fate of chemical and biological weapons, defence budget cuts in the wake of rearmament for the Korean War did.

During this period, statements of policy, committee discussions and expert assessments of the threat reflect ambiguities and uncertainties, with decisions on matters such as overall defence priorities and key equipment being reversed several times (Balmer 2001, Balmer and McLeish 2012, McLeish 1997). With respect to biological warfare, this drift failed to culminate in clear decisions at Cabinet level. By 1954, the Cabinet Defence Committee accorded a higher priority to defensive over offensive biological warfare preparations, while maintaining a need for a retaliatory capability. The Air Staff Target for an anti-personnel biological weapon was cancelled in 1954. Four years later, the Chiefs of Staff declared biological warfare to be an 'insignificant' deterrent. Chemical weapons fared little better. In July 1956, faced with defence spending cutbacks, Cabinet decided to abandon the large scale production of nerve gas and the development of nerve gas weapons and to destroy the residue of the Second World War stockpile of other chemical agents and weapons. A later decision was taken in 1957 to abandon the entire stockpile of chemical agents.

The turn to a defensive policy was re-visited during the early 1960s for a variety of reasons, including new scientific developments, changes in strategic thinking about nuclear weapons, intelligence on Soviet capabilities and changing NATO policy (Balmer 2001, 2010). In May 1963, the MacMillan Cabinet Defence Committee approved the development of a limited chemical retaliatory capability.[7] The committee also approved new research on offensive aspects of biological warfare but without the requirement for the UK to acquire a capability. These decisions came with funding over five years for work that included an increase in research and development on lethal and incapacitating chemical agents and their dissemination, production of a lethal chemical agent, production of defensive equipment and research into biological warfare agents, including large-scale trials.[8] Although plans were laid for a chemical retaliatory capability, which was to consist largely of spray tanks attached to aeroplanes and filled with the nerve agent VX, these plans were never realised. The Wilson Labour government, elected in 1964, did not abandon the policy of the previous government, but implementation was continually deferred and eventually entered into permanent abeyance, largely because the government's attention was diverted by economic crisis and a consequent reduction in spending at the Ministry of Defence (Balmer 2010).

7 TNA, CAB 131/28.D(63)3. Cabinet Defence Committee Meeting. 3 May 1963.

8 TNA, DEFE 11/660. D(3)14. Cabinet Defence Committee. Biological and Chemical Warfare Policy. Memorandum by the Minister of Defence, 16 April 1963.

By the late 1960s, chemical and biological disarmament had also come back onto the international agenda. While the topic had never completely gone away, the momentum for arms control was increased by breakthroughs in nuclear arms limitation, namely the Limited Test Ban Treaty (1963) and later the Nuclear Non-Proliferation Treaty (1968). Beyond government, various non-governmental organisations, in particular the Pugwash Conferences on Science and World Affairs, maintained interest in CBW disarmament. Then as the world became aware of the use of tear gas by the US in Vietnam, largely through media reports from 1965 onwards, the issue of chemical weapons became particularly sensitive in public. This was exacerbated because the USA was one of the only major powers to have signed, but not ratified, the 1925 Geneva Protocol. Following from a 1966 Hungarian proposal to the United Nations (UN) calling for all states (but especially the USA) to comply with the 1925 Geneva Protocol and declaring the use of CBW an international crime, various discussions took place both within and beyond the UN (Wright 2002, Chevrier 2006). The stakes were raised after a startling announcement by President Richard Nixon in 1969 that the USA would unilaterally disarm its biological and then toxin weapons (Tucker and Mahan 2009). In the course of negotiations a key sticking point had been over whether to negotiate a ban on both chemical and biological weapons. A Soviet draft proposal came in March 1971, turning back on its insistence from only a few months previously, of a singular chemical and biological weapons treaty, thus paving the way for the USA to negotiate on the basis of this text. On 10 April 1972 the Biological Weapons Convention was opened for signature, effectively outlawing this entire class of weapons. It took until April 1997 for a similar agreement on chemical warfare, the Chemical Weapons Convention, to enter into force and therefore become legally binding on signatory nations.

## Science and Secrecy

How do we answer a deceptively simple question about secret science: who knows what, where? Historian, Michael Dennis, went directly to the point when he wrote that 'far from being straightforward, the relationship of secrecy with the production of knowledge opens up a hermeneutic can of worms that science and technology studies must address' (Dennis 2007: 181). This section explores some of the literature that has addressed Dennis' 'can of worms'. It does not purport to be an exhaustive review of writing on secrecy, but rather a focussed attempt to make some initial comments about secrecy – particularly in relation to science – before engaging with the case studies in later chapters. We can start with Fox Keller's observation that the aim of science has often been articulated as the unveiling of the secrets of nature (Fox Keller 1993). This work of uncovering secrets, she notes, has often itself been performed in secret and possibly nowhere more so than in the concerted Second World War effort to build the first atomic bomb, the Manhattan Project.[9]

Indeed, military science, secrecy and transparency are most readily associated with nuclear weapons research (Badash 2003). After the atomic bombing of Hiroshima, a White House press release referred to the Manhattan Project as 'the greatest achievement of organised science in history' (cited in Hughes 2002: 93). It was, in fact, an almost unprecedented organisation of not only scientists, but also industry and the military. Moreover, a significant feature that accounts for the success of the Manhattan Project is the preoccupation with secrecy at the various sites involved in creating the atomic bomb. Compartmentalisation, telling people information on a strict need-to-know basis, meant only a few people had a complete overview of the project, which helped ensure both security and, in the words of the military director of the project, Leslie Groves, that the scientists would 'stick to their knitting' (cited in Thorpe 2007: 100). In this manner, efficiency, security, bureaucracy and secrecy all came together at once.

Matters involving biological and chemical warfare were also shrouded in secrecy. A number of recent historical surveys of biological and chemical warfare research have emphasised that all state sponsored programmes, together with sub-state activities, were cloaked in utmost secrecy (Guillemin 2005, Tucker 2007, Wheelis et al. 2006). Beyond the physical security of locks, guards and gates, there are many other means of achieving secrecy. In the case of the former Soviet Union, entire cities might disappear from the map and be closed off in order to conceal their military activities (Gentile 2004, Hart 2006, McCleish 2009). For instance, in McLeish's account of a rare post-Cold War visit to the closed city of Stepnogorsk, Kazakhstan, she records how secrecy operated on different scales simultaneously to obscure the biological weapons research taking place there (McLeish 2009). Work was embedded in a network of civilian microbiology research establishments, thus concealing it at an international-geopolitical level from satellite surveillance; a series of cover stories and 'legends' were devised and circulated to prevent the spread of the secrets within the Soviet Union; and finally, at the local level, the physical separation of the civilian and military zones of the compound was reinforced by a policy of not employing workers from the civilian arm in the military section.

It is fairly straightforward to suggest reasons for this sort of secrecy in matters relating to military science. It keeps enemies out. Practices and arrangements for secrecy guarded the knowledge of often untried and potentially devastating weapons, and also shielded the policy-making process. But secrecy also had other effects. A degree of isolation separated workers from other communities, scientists working in secret military organisations were often restricted in activities such

9 Keller also explores the relationship between gender, secrecy and knowledge. A similar theme is explored in relation to medieval anatomy by Park (2006), whereas, in relation to chemical weapons, the often asserted association between women and poisoning is discussed critically in Kord's chapter 'The Female Self: Posioners' (Kord 2009: 154–186).

as publication, conference attendance and patenting, thus largely divorcing them from the reward system and other social institutions of open academic science. Secrecy and security also served to isolate scientists from direct interaction with groups or individuals opposed to their work on moral grounds; this is a topic which we will return to in future chapters. Regardless of existing protest against this work, secrecy could potentially contain any further spread of disquiet, acting to prevent further revelations that might lead to even wider public censure. Finally, at least for countries that had ratified the 1925 Geneva Protocol, secrecy surrounding chemical and biological warfare would have minimised the possibility of a nation being accused, whether accurately or inaccurately, of breaching international legal obligations.

As well as suggesting reasons behind secrecy, it is also possible to ask normative questions about the desirability and effects of secrecy: how secrecy works and what secrecy permits. So, in his work on secrecy in the McCarthy era, sociologist Edward Shils expressed his concern about the corrosion of democracy when secrecy, privacy and publicity are not in equilibrium (Shils 1956, Blank 2009). Bok's work on the philosophy of secrecy is subtitled 'on the ethics of concealment and revelation' and places normative issues at the forefront of her analysis (Bok 1989). Turning specifically to knowledge production, when commentators treat secrecy in science as more than simply part of the background, it is often taken as a conceptually unproblematic condition under which scientists sometimes labour, even if this gives rise to interesting or ethically problematic dilemmas over issues such as data-sharing or commercialisation (e.g. La Follette, 1985). And in military, commercial or academic science, secrecy is often treated as inherently pathological and contrary to the professional ethos of science (Chalk, 1985, Kenney, 1986, Resnik, 1998).

Accounts that take it for granted that secrecy is bad for science make a tacit nod towards sociologist Robert Merton, who argued that various norms of behaviour are internalised during the training of a scientist and enforced through the reward system in science (Merton, 1973). One such norm, communalism, suggests that scientists will learn to share their results through open publication and be rewarded through recognition of their achievements. As the antithesis of communalism, secrets are taken to be inherently bad for science. But this account of science is problematic, resting on a precarious history of science. Modern science has always been a subject for concealing. This is nowhere more apparent than in Bacon's fictional but visionary account of the House of Soloman. In his 1624 *New Atlantis*, Bacon described the House of Soloman as an organisation dedicated to experiment and discovery, but also to a degree of secrecy:

> And this we do also: we have consultations, which of the inventions and experiences which we have discovered shall be published, and which not: and take all an oath of secrecy, for the concealing of those which we think fit to keep secret: though some of those we do reveal sometimes to the state and some not. (Bacon 2006 [1624])

Later in the century, we find Isaac Newton keeping his alchemical work secret (Golinksi 1988). At first sight this might appear to be because he made a strong epistemological divide between his 'scientific' chemistry and 'unscientific' alchemy and theology. But, as most contemporary historians of science would concur, this is anachronistic. For Newton, all three were important and interwoven. Alchemy, though, promised 'sacred and powerful truths' that could imitate God's activity at the Creation. No wonder, argues Golinski, that it was imperative for Newton to keep this potentially dangerous knowledge restricted to an elect few.

Merton's account of openness, as one of the norms enforced through 'the' reward system and one that provides a sociological underwriting for good science, is rendered even more problematic by empirical evidence about how scientists can inhabit multiple reward systems. For example, in the seventeenth century Galileo quite happily sent his telescopes to princes and cardinals, who would support his work, but refused potential competitors or critics (most famously Kepler) the same service (Biagioli 2006). Even when they ostensibly work within the same system of rewards, scientists may claim that secrecy is good for science. Mulkay, in a direct attack on Merton's theory, showed how the group of Cambridge astronomers who discovered pulsars in the late 1960s defended their practice of withholding information against critics who accused them of undue secrecy; this defence was on the grounds that they needed to check their data and protect their more junior researchers against competition (Mulkay 1976). Mulkay's more general point is that norms – including communalism – neither describe nor prescribe scientific behaviour in its entirety, but instead can be drawn on as a flexible 'moral vocabulary' to praise or condemn a scientists' own behaviour or the behaviour of other scientists.

The idea that secrecy is not automatically antithetical to scientific practice should not suggest that a case cannot be made to condemn secrecy under certain conditions. Commentators frequently note how readily the secrecy introduced from a genuine need for security elides into the use of secrecy to mislead and protect ineptitude, malpractice and other potential embarrassments (Gibbs 1995). These two types of secrecy correspond closely to what Shils called functional secrecy, stemming from 'practical necessity', and symbolic secrecy, which Shils describes also as paranoid and 'maximal-loyalty' security (Shils 1956: 235–6).

Loss of accountability entailed by secrecy is frequently raised as another telling criticism of undue secrecy, particularly in discussions about military-related science. Guillemin, in this vein, makes a forceful argument that secrecy has at times allowed biological warfare scientists to cut loose from a 'moral compass' (Guillemin 2005, 2006). This loss had practical effects, for example at the end of the Second World War, when it stifled possible public debate on the continuation of war-time biological warfare research into peacetime. Referring also to the US authorities' exchange of data – including data from humans subjected to biological weapons tests – from the Japanese biological warfare programme, in return for immunity from prosecution for its head General Shiro Ishii (Harris 1994), Guillemin argues that 'by keeping the dreadful secrets of the Japanese biological warfare program, the US government lent legitimacy to offensive biological warfare aimed against civilian populations' (Guillemin 2006: 91).

Rather than prejudge the value of secrets, philosopher Sissela Bok adopts a working definition of secrecy as 'deliberate concealment' (Bok, 1989). Along similar lines, Perkins and Dodge note that 'a commonly accepted definition of secrecy sees it as the practice of selectively sharing information, but at the same time hiding it from other groups' (Perkins and Dodge 2009). These, relatively neutral, definitions avoid forging an immediate moral link between concealment and shame and instead open up a space to debate the ethics of secrecy as it occurs in different circumstances. Additionally, by construing secrets as beneficial to some and not others, we can then ask whose interests any particular clandestine arrangement serves. This approach may then arrive at normative judgements about secrecy, yet does not automatically condemn or endorse it out of hand.

A second advantage to Bok's notion of 'deliberate concealment' is that it points us to the active nature of secrecy: someone does the concealing. So, beyond normative questions, it is useful to think about how secrecy is enacted and maintained in practice, how it becomes a 'condition in which action takes place' (Carmeli and Birenbaum-Carmeli, 2000). A fairly common way of viewing secrecy is in terms of control by those in power over flows of information (e.g. Agar and Balmer, 1998, Cloud and Clarke, 1999, Moynihan 1999). In this vein, Weber's classic works on bureaucracy point to secrecy as central to the wielding of bureaucratic power (Weber 1920 [2009]). As Blank comments, 'for Weber a secret is simply knowledge that someone else lacks', with the bureaucrat able to exploit being on the surplus side of this differential (Blank 2009). Along similar lines, Simmel in his 1906 seminal essay on secrecy, regarded concealment as fundamental to social life, a crucial way for actors to deliberately shift the balance between two crucial elements of social interaction – what is known about another person, and what has to be assumed (Simmel 1906).

So, information, together with exchange or withholding that information, is a useful starting point in thinking about the mechanics of secrecy.[10] One disadvantage of this perspective, however, is that it can lead to secret information being treated as a single thing, rather than as a more complex series of social arrangements. Bratich has condemned this as an 'obsession with secrecy as a box to be opened' (Bratich 2006: 494). This is not only a problem for analysts. Social actors can readily adopt the view that 'the secret' is a contained and bounded entity. Weart, for instance, has observed that in the late 1940s the secret of the atomic bomb, singular, was pervasive in cultural representations of nuclear matters, 'as if it were a formula on a piece of paper in a safe somewhere' (Weart 1988: 121, see also Kaiser 2005).

---

10 There are resonances here that I will not pursue with the vast literature on cooperation that has grown in game theory around the 'prisoner's dilemma' and the concealment or revelation of items of information (Axelrod and Hamilton 1981).

Besides the idea that secrecy is construed as the contents of a box, envelope, safe or other container, Bratich draws on Deleuze and Guattari's discussion of secrets in *A Thousand Plateaus* to add two other ways of re-thinking secrecy (Bratich 2006, Deleuze and Guattari 1987). Resonating with the dynamic sense of 'deliberate secrets' just discussed, Bratich adds a second component of secrecy, the influence and effects of secrecy, which can themselves also be hidden. In this respect, think of how conspiracy theories draw on this sense of secrecy to suggest the existence of a multitude of clandestine influences on society (Parish and Parker 2001, Knight 2000). Thirdly, secrecy also generates secret perceptions through the acts of spies and blackmailers. In this last sense of the term, as sensational revelations are threatened or divulged, the boundaries between secrecy and privacy blur. A personal private letter, for instance, falls into the wrong hands and becomes an occasion for someone to threaten to reveal secrets that could ruin someone's reputation or career (Welsh 1985).[11] Dennis captures this aspect of secrecy when he writes 'Secrets are only known when they are no longer secrets, but the power to unveil and display a secret is what makes secrets useful and dangerous' (Dennis 2007: 178).

Simmel, Derrida and others have also drawn attention to revelation as a threshold between openness and secrecy. For Simmel (1906), the appeal of concealment and the opposing need to reveal secrets, whether through temptation or pragmatic decision, exist in a tension that is heightened and released at the very moment of revelation. It is also, according to Simmel, often at this point – at the moment of its disappearance – when the value of the secret is most intensely recognised. Cultural theorist Jeremy Gilbert makes an adjunct point (Gilbert 2007). In his comments on Derrida's appeal to secrecy as something consciously kept secret (Derrida and Ferraris 2001), and admitting that elsewhere Derrida implies that absolute secrecy is something that could never be known by others and therefore does not need keeping in any active sense, Gilbert suggests that 'we might take this further and ask if a secret is really a secret as-such at all before it has been told. What if it is only the act of disclosure, surveillance or confession which constitutes any particular piece of the continuum of experience as "a secret"?'. On all these points, I would suggest an alternative. As we will see later in this book, once a secret is revealed it frequently engenders a debate (whether within the secret-knowing community or with the new audience for the secret) about whether it ever really was a secret in the first place. From the case studies, it appears that breaches of secrecy provoke debate about what exactly was secret, what exactly was disclosed, and what exactly the consequences of disclosure might be. It is in

11 Helpful discussions of the differences between secrecy and privacy can be found in Vincent (1998: 18–25) and Bellman (1981). Bellman notes that conventionally, 'The term private usually establishes that the other person does not have the right to some knowledge… A secret, on the other hand, concerns information that the other person may have rights to, but that the possessor chooses, is told to, or is obligated to withhold'(Bellman 1981:4).

this sense that the moment of revelation shows 'the secret' as unstable. The very labelling of a secret is contestable.

Secret knowledge is therefore not just an arcane thing sitting in a safe or in the minds of scientists. Other literature in the sociology and anthropology of science also treats secrecy as actively produced and maintained. As Wright and Wallace argue, 'secrets do not develop in a social vacuum. Rather, the construction of a web of secrecy is a social process that defines relationships between those inside and outside the web.' (Wright and Wallace 2002: 369).

Equally, commitment to secrecy entails commitment to many ancillary 'complexes of behaviour' (Shulinder 1999: 89). In a nuclear weapons laboratory, for instance, secrecy dictates where people can go and what they can do, depending on their security clearances (Gusterson 1996). Other ancillary behaviour can be more elaborate, such as when Cold War US scientists working on military projects accommodated national security by reproducing their own version of the open scientific community, with their own peculiar array of closed conferences, publications and committees (Westwick 2000).

Deliberate concealment of knowledge is therefore grounded in the mundane: documents, places, routines and suchlike. Geographical perspectives in particular help to bring this aspect of secrecy into view, notably work that focuses on the geography of science (e.g. Gieryn 2000, Naylor 2005, Powell 2007, Shapin 1998). As Paglen puts this point, in contrast to the abstraction of secrecy in debates about democracy or politics, 'thinking about secrecy in terms of concrete spaces and practices helps us to see how secrecy happens and helps to explain how secrecy grows and expands' (Paglen 2009: 16). Considering secrecy in spatial terms points to the fracturing of space into regions of knowledge and ignorance. Spaces are produced through 'technologies of privacy' (Hilgartner, 2000), ranging from mundane elements such as locked doors, filing cabinets, badges and access privileges, through to techniques of control, such as compartmentalisation and classification, that embrace entire organisations (Gusterson 1996, Hilgartner 2000, Reppy 1999). Such spaces are also marked by less intentional concealment, what Vaughan, in her analysis of the 1986 Challenger Space Shuttle disaster, calls 'structural secrecy', where the same terms (such as 'anomaly' or 'hazard') were used and interpreted differently in different parts of the NASA organisation (Vaughan 1997).

Secrecy and classified knowledge have also been conceptualised as anti-epistemology (Galison, 2004). Galison explains that 'epistemology asks how knowledge can be uncovered and secured. Anti-epistemology asks how knowledge can be covered and obscured' (Galison 2004: 237). The term is provocative, although it is in danger of running counter to the thrust of the geographical arguments just considered. It implies that there are just two homogenous worlds, an outside where everyone knows public knowledge and a hidden world where nobody knows secret knowledge. Yet, as Galison points out, a rough calculation places the volume of restricted information in the closed and classified world far higher than that of public information. And somebody, somewhere, does have access to this information. To them, it is anything but anti-epistemological. The majority of codified knowledge might be obscured, but not to everyone. Anti-epistemology, I will argue later in this book, does not so much deny knowledge, as it fractures and disrupts the topography of knowledge – providing particular geographically restricted accounts of the world. In this sense, secrecy acts as a spatial-epistemic tool in the exercise of power.

So, to pull several threads of this discussion together, while at first glance Simmel's often cited remark that 'the secret offers… the possibility of a second world alongside the manifest world' (Simmel 1950 [1906]) does capture a sense of insider and outsider perspectives, it perhaps suggests too strongly the separation of these two worlds. As discussed, covering and uncovering do not happen cleanly and automatically but are thoroughly social, spatial and political processes. Secrecy is enacted, or performed, as the 'technologies of privacy' actively and continually contribute to the production of multiple worlds.[12] Such multiple worlds are circumscribed but not entirely isolated. Even where the secret is quite literally a thing in a box, the hidden and revealed maintain connections. Empson draws on this insight in her anthropological analysis of a box in a Mongolian household where kinship relations invoked by the items (such as photographs) placed on display, and those invoked by items kept hidden in the box (such pieces of umbilical cord) depend on each other for their meanings and significance (Empson 2007). More generally, for Empson, openness and secrecy should be regarded not as either/or but as foregrounded and backgrounded in any situation.

Taking similar ideas about the connectedness of secret and open spaces back into a military setting, Dennis compares civil society to an archipelago, with secrecy as an ocean obscuring the military institutions that connect the islands (Dennis, 1994). His metaphor illustrates that 'public' and 'secret' have submerged connections, made through such acts as assigning security classifications that allow some outside people or groups 'in' while keeping others out. Zones of interaction have existed between nominally separate civilian and military institutions, such as industrial firms and military research laboratories (Cloud, 2001). As such, the boundaries between secrecy and openness are fluid and negotiable. Furthermore, as we will see in the remainder of the book, as knowledge moves from the secret spaces of military knowledge production, through the media and through different government departments, radically different accounts of what is known and not known can be articulated in each arena.

---

12 Enactment of multiple identities and objects has recently been explored in a Science and Technology Studies context by Mol (2002). The possibilities of the 'same' text being read as multiple objects in different contexts has been explored by various authors, notably Mulkay (1989).

## Sources and Historiography

From a historiographical perspective, it is notable that while the material in the following chapters presents a more detailed narrative than any previously available, it is nonetheless assembled almost entirely from documentary records. These are, again almost entirely, official records released at the National Archives, Kew, London. Many of the documents were originally classified as 'secret' or 'top secret', a point to which we will return in Chapter 4. The policies, plans, reports, memoranda and minutes in these documents were all written and discussed within the pervasive culture of secrecy in Whitehall (Hennessey 2001, 2002, Vincent 1998). This culture was overshadowed by a series of Official Secrets Acts, most notably the 1911 Official Secrets Act, which made it illegal for civil servants to communicate official information to the outside world.

While often surprisingly detailed, these official documents must be read and interpreted as official sources, frequently performing a rhetorical function within Whitehall, such as recording 'in stone' the final outcome of a more submerged and uncertain process of negotiation and decision-making, or trying to persuade colleagues of a point of view or course of action. So, on a reflexive note, it is important to be explicit that my account of secrecy itself constructs transparency. In other words, I have pieced together this account from a huge number of documents, many of which were closed until recently and not written specifically to create a historical record, into a narrative that reads as if it is an open window onto past hidden events. Such openness depends on the 'public secret' – something generally known but not readily spoken about – suffusing my narrative (Taussig 1999). Here, the public secret is simply that the transparency of my account is (of course) accomplished, a 'black-boxed' effect of my interpretive wrestling with primary documents and few first-hand accounts to produce a seamless and relatively coherent history.

## Overview

The next chapter tackles the ethics of secrecy in science related to the military primarily from a historical and sociological, rather than philosophical perspective. After briefly considering the normative arguments for and against scientists undertaking research in this field, it turns to an empirical exploration of how past British scientists, principally scientific advisors, attempted to defend research on biological weapons. Although the historical record is scant, sufficient examples exist to detect a degree of continuity in their justifications, and a number of themes can be identified. Scientists argued: that biological weapons research is morally justified because it produces humane weapons; that it is no different from medical or other research; and that it was being performed for defensive purposes. Although these arguments appear to be directed at putative objectors outside the research and policy-making arenas, I argue instead that these defences were caged in by secrecy and directed primarily towards other scientists working on germ warfare, thus forming a significant component of the 'moral economy' of that secret community.

Chapter 3 explores secrecy in the practices of military research through a case study of an accident at a secret biological weapons field trial. An integral part of UK Cold War research on biological warfare involved a series of outdoor trials at sea where scientists exposed animals to clouds of pathogenic micro-organisms. Not all of these trials ran according to plan and at the culmination of 'Operation Cauldron', a series of trials that took place between May and September 1952, a fishing vessel, the *Carella*, strayed into the danger zone around the trial shortly after a test bomb had been detonated. The *Carella* ignored warnings and the ship's crew was exposed to an invisible cloud of germs. The government's response was to instruct the navy to tail the trawler over the next month without notifying its crew, and to listen for a distress call. This chapter explores the decision-making process during the incident and the wider implications for our understanding of science and secrecy. My aim is to highlight not simply that these events occurred in utmost secrecy, where it became a guiding principle that overrode other considerations, but equally to demonstrate how secrecy operates and what secrecy allows. Two processes are apparent for maintaining secrecy. First, actors treat the secret as a self-contained thing, a singular entity. Second, those who hold the secret attribute a tremendous amount of knowledge and insight to those people outside the culture of secrecy. Furthermore, the existence of secrecy permits familiar categories to blur, in particular those of accident/experiment and patron/scientist.

The following chapter, 'Keeping, Disclosing and Breaching Secrets', focuses on security classification as a fundamental dimension of secrecy. I will look at two sides to the operation of secrecy: keeping secrets in and letting secrets out, with a focus on how secrecy and transparency are operationalised through security classifications (i.e. Top Secret, Secret etc). A close look at debates about how and what to classify as a secret, and decisions as to just what degree of secrecy to apply to information about biological and chemical warfare, reveals a series of choices that reflect wider concerns, in particular co-ordination and co-operation between national research programmes. Turning next to how secrets escape, I will discuss how breaches of secrecy were dealt with in the history of the UK programme. In this section, I want to contrast the controlled release of secrets through press releases and other announcements with the uncontrolled breach of secrecy exemplified by leaks. Restoring control becomes a matter of the authorities employing various resources, including resources such as rumour and gossip, normally seen as the province of those to whom secrets are leaked.

Chapter 5 examines the converse side of the relationship between secrecy and knowledge production; it is about secrecy, uncertainty and ignorance. My main point is that scientists acting as expert advisors behind closed doors could draw on their cultural authority as experts not only to lay claim to certainty, but also to uncertainty. Put bluntly, scientists in this context became socially legitimated doubters. The empirical focus is on expert advisors in the history of the British biological warfare programme after the Second World War. During this period, as Britain moved from an offensive to a defensive stance over biological warfare, the status of biological warfare as an opportunity and threat, the status of research on biological warfare, and the significance of scientific advisors in the policy process all varied. Much of the time, the expert advisors presented information on biological warfare to their military and political audiences as factual information, which spoke for itself with straightforward, direct and unequivocal consequences for policy. The status of the advisors and their scientific constituency was thus reinforced. Under conditions of secrecy, the same advisors, however, were quite capable of weakening their own claims without being prompted by any adversarial challenge. Experts frequently portrayed the findings of biological warfare research to their audiences as uncertain, in dispute and with negotiable consequences for policy. When successfully deployed, the effect, rather than diminishing the status and autonomy of the experts and their constituent scientists, was to present the field as one demanding more research, more resources and a higher status in policy deliberations.

The following chapter directly explores the relationship between secrecy and openness, developing ideas introduced in the previous chapters. In 1967 and 1968 the chemical and biological warfare research establishments at Porton Down, Wiltshire were the subject of a number of television programmes. Although the 1967 programmes adopted a fairly deferential attitude towards Porton Down, by 1968 the coverage was openly hostile. This portrayal was accompanied by a series of protests, parliamentary questions and other unwanted publicity. In this chapter, I describe and discuss how Porton Down and the Ministry of Defence reacted and responded to this spate of poor publicity, as officials moved from simply expressing indignation to an organised campaign of public relations management. The chapter will explore how secrecy and transparency are inter-related: particular types of transparency were enacted by the authorities through the careful management of what information was released and what was kept secret. Throughout this chapter, the common metaphor of a sphere of secrecy does not do justice to the operation of secrecy in practice. Degrees of openness and secrecy, choices over what to reveal and conceal, point to a geographical analogy, that openness exists as a zone between two others: the initial secret, then what is revealed about that secret, but also what is held back and remains out of sight.

The final empirical chapter explores how a geography of knowledge can be developed around the idea that secrecy, space and knowledge are inter-related. It focuses on the publication and subsequent treatment in 1975 of a newspaper article reporting that the patent on the chemical warfare agent, VX, was available in a number of public libraries. Within ten days, copies of the patent had been withdrawn, a government review of declassification procedures was announced, and in Parliament the Minister for Defence announced that the Government had never patented VX. The implication was that nothing, or nothing worth worrying about, had happened. The chapter traces the modifications of position that occurred in order for the Minister to arrive at this announcement. I argue that secrecy enabled different readings of the patent in different places and thus acted as a 'spatial-epistemic' tool in the exercise of power. Key features that differed were: the relationship between essential properties attributed to VX and the additional tacit knowledge deemed necessary to make the nerve agent; the degree of revelation that was deemed to have occurred as the secret was differently constructed; and the presumed intent and abilities of putative abusers. Building further on arguments from earlier in the book, we see how secrecy actively constructs knowledge (such as what is or is not dangerous, what is or is not hidden etc) as well as ignorance (such as the idea that nothing of importance happened). By this stage, we will have travelled a long distance from the opening observation that it is too simplistic to claim that secret science is the same as open science, just done behind closed doors.

# Chapter 2
# Secrecy at Work: Scientists' Defence of Biological Warfare Research

Arguments about the moral status of biological warfare and research are well documented. The characteristics that mark out biological warfare for special moral opprobrium are: its uncontrollable and indiscriminate effects, its insidious nature, its deliberate perversion of medical science, and its potential to threaten our survival as a species (Adams 1986, Haldane 1987, Sims 1987, Royal Society 2000). This notion of 'public health in reverse' has also been used to condemn scientific research on biological warfare (Lappé 1990, Maclean 1992, Sinsheimer 1990), often coupled with the observation that it is extremely difficult morally to separate offensive from defensive research. An alternative strand of analysis, which has been applied rigorously to the chemical warfare taboo, argues that it is mistaken to seek out the roots of the moral status of chemical or biological weapons in any of their essential characteristics (Price 1995, 1997). The moral status of these weapons at any particular time must, according to this argument, be regarded as the contingent outcome of power struggles between competing political discourses. At stake in the competition is what will count as the truth about the morality of chemical and biological warfare. While these two approaches – essentialist and contextualist – differ on the universality of the chemical and biological weapons taboos, neither analysis denies that they can possess force or reality. More recent discussion extends this analysis to biological warfare and poses a dual challenge to both Price's 'anti-essentialist' perspective and also to its opposite position, that there are inherent characteristics of CBW that make them taboo (Jefferson, 2009).

Regardless of the roots of the taboo and fear surrounding biological and chemical weapons, if the norms against biological warfare are so evident, what has motivated scientists historically to participate in biological warfare research? According to Colwell and Zilinskas (2000) the possibilities include scientific interest, intellectual challenge, better remuneration, job security, national security and, in some programmes, explicit and implicit threats to the individual. While all of these explanations are plausible, they imply that scientists have not participated in biological warfare research as a matter of course, and instead that exceptional factors are needed to spur scientists into violating the norms of the scientific community. As discussed in the previous chapter, whether or not the normative structure of science has such a determinate effect on the behaviour of scientists has been much debated within sociology of science. Briefly, two schools of thought prevail (Merton 1973, Mulkay 1976). In the classical view, norms such as universality, communality, disinterestedness and organised scepticism describe an empirical reality within the scientific community, with rewards and sanctions in place to ensure conformity. More recently, sociologists have argued that norms are not related to the scientific reward system. Scientists who apparently flout the norms may still receive recognition for their science. Norms instead function as part of a professional ideology and moral vocabulary that paints a positive image of science. This image can be used internally, to defend or condemn colleagues' behaviour, or externally, to help secure popularity or resources for the profession.

Adopting a strictly classical view suggests that the weapons scientist enters some sort of moral vacuum by stepping into secret laboratories and outside the accepted normative structure of science. Cole (1996: 217–19) has similarly suggested that the secrecy in the US biological warfare programme has been less about security, and more about the desire to shield research from moral opprobrium. And, more generally, in post-Second World war popular culture weapons scientists have frequently been portrayed less as evil or mad scientists and more as amoral idealists pursuing pure science without regard to its consequences (Haynes 1994: 246–7).

Yet, scientists working within national biological warfare programmes certainly found ways to justify their work. A parallel can be made with nuclear weapons scientists and allied experts, where both outside observers and past participants have noted how these communities adopt abstracted language that distances their work from the sheer horror of mass killing (Cohn 1987, Nash 1980). So, as anthropologist Hugh Gusterson points out in his closely observed ethnographic study of weapons scientists at the Lawrence Livermore National Laboratory: '...rather than ignore the ethical dilemmas of their work… [they] learn to resolve these dilemmas in particular socially patterned ways. In other words, becoming a weapons scientist involves much more complex and creative social and psychological processes than repression and avoidance' (Gusterson 1996:42).

Understanding these complex and creative processes with respect to biological warfare scientists requires insight into the constitution of the moral economy of the biological warfare research community. Moral economy is a term adapted by historian Robert Kohler to describe how, in laboratories, the 'unstated moral rules define the mutual expectations and obligations of the various participants in the production process' (Kohler 1994: 11–13). For Kohler, moral economies are localised, rather than global across the whole of science, and are tied to particular material and social arrangements. Furthermore, the flexible appeal to norms demonstrated by critics of classical sociology of science suggests that these expectations may at best be aspirational, and will certainly form part of a professional ideology, but will not dictate how scientists behave. From this perspective, there is no reason to expect the moral economy of the military laboratory to resemble that of the university or commercial worlds.

Picking up from these observations about the complexity of the moral economy of science, this chapter presents historical cases in which British scientists associated with biological weapons research, primarily scientific advisors, attempted to defend their work. There is a degree of continuity in their justifications for biological warfare research, and a number of themes can be identified. Biological warfare research was justified because its products were claimed to be at least as, if not more, humane than conventional weapons; at times it was lauded for producing civilian applications, particularly in medicine; a final set of themes invoke patriotism and defence of the country as warranting the pursuit of biological weapons research. A second aim of this chapter, more tentative given the scant evidence available, is to ask the question how did these arguments made sense to their proponents? Here, I regard secrecy as of paramount importance largely because, in principle, it should remove any requirement to articulate justifications for ones' actions. With this in mind, I suggest, and spell out later, that the various strands of justification only make sense as part of the moral economy of the weapons laboratory, and as such are told for the benefit more of the tellers than for any external audience.

Although I discussed general historiographical problems with sources in the first chapter, it is worth adding here that government correspondence and committees are hardly the place for lengthy discussions of the ethics of biological warfare research, and there are only a few instances on record where scientists made direct reference to the morality of biological warfare research. These records contain the views of elite advisors, rather than of scientists engaged in day-to-day research. My discussion also tries to avoid repeating the careful historical work done by Schmidt (2006, 2007) and Evans (2001) on the history of human experiments with chemical weapons at Porton, both of whom address wider questions of informed consent and medical ethics. My focus instead is on what scientists say about the value of their own research in a culture where killing people is the goal of the work.

## Early Justification of Biological Warfare

The prospect of what was then termed bacteriological warfare remained almost entirely speculative before the First World War. To be sure, crude forms of biological warfare date back to antiquity (Mayor 2003). In science fiction, 'The Stolen Bacillus' appeared in H.G. Wells' first book of short stories, published in 1895, in which an anarchist steals a sample of cholera bacilli from a bacteriologist. The anarchist plans mayhem by spreading disease throughout London, but is thwarted when the bacillus turns out not to be cholera, but a microbe that turns its victims a shade of blue. While stories such as this show remarkable prescience, it was still the prospect, rather than the reality, of biological warfare that prompted its ban under the 1925 Geneva Protocol. Germany had attempted sabotage operations with biological agents during the First World War, but these clandestine operations remained hidden until well into the twentieth century (Wheelis 1999). Instead, it was widespread revulsion towards chemical weapons used in the war that provided impetus to the protocol, with bacteriological warfare added as an afterthought (McElroy 1991). Not everyone concurred with the idea that chemical weapons were morally repugnant (see Holden Reid 1998, Spiers

2006). Most notably among practicing scientists, biologist JBS Haldane made a public defence of chemical warfare in his 1925 short book *Callinicus*. A major strand of Haldane's argument was that there was little difference between gas and conventional weapons:

> They [applications of science] can all, I think, be abused, but none perhaps is always evil; and many, like mustard gas, when we have got over our first not very rational objection to them, turn out to be, on the whole, good. If it is right for me to fight my enemy with a sword, it is right for me to fight him with mustard gas. (Haldane 1925:82)

It was a newspaper report suggesting that German secret agents had infiltrated the London Underground in order to carry out experiments with bacteria that prompted the British government to seek expert advice on the potential of biological warfare from a small number of civilian scientists consulted in secret. The most influential were Professor John Ledingham, Director of the Lister Institute; the epidemiologist and bacteriologist Professor William Topley of the London School of Hygiene and Tropical Medicine; and Captain Stewart Ranken Douglas, Deputy Director of the National Institute for Medical Research. Their 1934 'Memorandum on Bacteriological Warfare' became a benchmark for the deliberations of the Committee of Imperial Defence's Sub-Committee on Bacteriological Warfare.[1] When the Sub-Committee first met in November 1936, the chairman and influential civil servant, Maurice (later Lord) Hankey, described the document as the 'datum line' for assessing the potential hazard.[2] While much of this report weighed up the possibility of germ warfare, and the forms it might take, the authors ventured briefly but significantly into ethical territory, and wrote:

> ...only opinions, usually exaggerated and ill-informed ones, have been expressed as to its probable employment in future wars... we take the opinion that this weapon cannot be ignored as a weapon of offence and that its 'ethical' standing is neither lower nor higher than that of the bayonet, the shell and the chemical arm. Indeed if its 'ethical' standing as a deliberate instrument of warfare is to be judged by the 'fighting chance' it gives its victim, it would in our opinion take a higher place than any of these three recognised instruments.[3]

1 TNA,W[ar] O[ffice] 188/648. Appended to CBW2. Committee of Imperial Defence Sub-Committee on Bacteriological Warfare. Note by the Joint Secretaries (4 November 1936). Memorandum on bacteriological warfare (13 April 1934).

2 TNA, WO188/648. Committee of Imperial Defence Sub-Committee on Bacteriological Warfare. 1st Meeting (17 November 1936).

3 TNA, WO188/648. Appended to CBW2. Committee of Imperial Defence Sub-Committee on Bacteriological Warfare. Note by the Joint Secretaries (4 November 1936). Memorandum on bacteriological warfare (13 April 1934).

At the outset, these scientific advisors were prepared to take a moral stance on biological warfare. In their opinion, the infectious nature of pathogens – one of the very features that has placed biological weapons apart from other forms of warfare – was to be counted as a virtue. This quotation remains the only direct defence, before the Second World War, apart from a fleeting reassurance by Hankey, at the inaugural meeting of the Sub-Committee on Bacteriological Warfare in 1936, 'that we were, of course, considering the problem from the defensive aspect; it was unthinkable that we should contemplate the adoption of bacteriological warfare offensively'.[4]

## Retaliation and War

Four years later, Hankey revisited this commitment. Hankey had become Minister without Portfolio, and his advisory committee had been reconstituted as the War Cabinet Bacteriological Warfare Committee. Under its auspices, Hankey instructed Professor Ledingham to prepare a research agenda in the field of biological warfare, stipulating:

> the programme should be drawn up on the assumption that there was no intention of taking the initiative in offensive measures, but should at the same time not exclude any experimental method, just because it might appear offensive in aim, if it seemed likely to [lead to] improvement in defence.[5]

The end of 1940 saw this plan put into action under the leadership of Paul Fildes at the new Biology Department Porton (BDP). Fildes had a strong reputation as a bacteriologist, and in the decade prior to the war had been instrumental in founding the sub-discipline of bacterial physiology through his work on bacterial nutrition (Kohler 1985). As head of the programme, Fildes was frequently in correspondence with Whitehall, although there are only a few references to his views on the ethics of his research. The first occurred during 1943 in relation to the in-coming chairman of the Bacteriological Warfare Committee. This position had been assigned to the Chancellor of the Duchy of Lancaster, Ernest Brown. Fildes complained:

> I am sorry to hear that Mr Ernest Brown is likely to be given the Chairmanship of the Committee. I have always understood that he is a strict Methodist, and that

4 TNA, WO188/648. Committee of Imperial Defence Sub-Committee on Bacteriological Warfare. 1st Meeting (17 November 1936).

5 TNA, WO188/653. BW(40)1st Meeting. War Cabinet. Bacteriological Warfare Committee (7 February 1940).

> seems to me to be a strange background for a person who has to support the less orthodox activities of the Committee.[6]

While recognising that there might be religious objections to biological warfare, Fildes did not expand on whether or not he believed this was a serious problem. By far the most substantial and revealing statement of his views came towards the end of the war, in a response to a report by a member of the Canadian General Staff, Colonel Goforth. The colonel claimed that the military on both sides would be deeply suspicious of both chemical and biological warfare, arguing that:

> From experience and study he [soldiers, seamen and airmen] has learned the capacities and limitations of old and new weapons involving various combinations of HE [High Explosive] and metal. He is instinctively suspicious of new methods of warfare such as CW [Chemical Warfare] and BW [Biological Warfare] may offer... they are NOT predictable or measurable in their military consequences. They tend to upset his careful staff tables and logistical calculations.... If toxic chemicals are regarded by him as an unsoldierly weapon, his much greater revulsion against BW is based on a possibly unreasonable belief that it is unclean and farthest removed from the weapons of chivalry.[7]

Goforth also added that widespread abhorrence of chemical warfare was evidence enough that the public would adopt a similar attitude towards biological warfare.

These arguments of unpredictability, lack of chivalry and public revulsion were countered by Fildes by comparing biological weapons with other forms of killing. He wrote:

> I do not think it true to say that a 'substantial minority' of the population would object to retaliation by gas, microbes or anything else, if an attack in these forms were made on them. They would be guided by a professional assessment of the military value of the retaliation.... Actually, any moral objections which may exist to BW are based on insecure grounds. Is it any more moral to kill Service men or civilians with HE than with BW? It may be agreed that it is not, but that BW is more 'horrific' in the sense of introducing unnecessary mental and physical suffering and so should be excluded if possible. The 'horrific' nature of BW is, however, not admitted by those who are competent to judge, on the evidence available from animals.[8]

6 TNA, WO188/654. Letter Fildes to Allen (27 November 1943).

7 TNA, WO 188/654. Review of the Most Available BW Agents Developed in Canada By C.1. Committee (31 July 1944). Appended Review of BW.GS Comments.

8 TNA, WO188/654.BIO/5293. Notes on Professor Murray's Memorandum (by Fildes) (7 September 1944). Goforth had also written his comments as a response to this memo outlining the state of biological warfare in Canada.

Fildes described how animals subjected to anthrax would inhale the spores, and not display any symptoms for a few days, but then die of septicaemia (blood poisoning) within a few hours, adding that 'observation of monkeys does not suggest that they undergo any more suffering than most people do when they are ill'. On the other hand, Fildes continued, a civilian caught in a bombardment of high explosives would undergo 'considerable mental disturbance long before he is buried under a pile of rubble'. This would continue, along with physical distress, upon removal to hospital where:

> the suffering he has undergone is often terminated by bacterial septicaemia similar to that which has the same effect as BW without the distressing preliminaries. It seems clear to me that a substantial majority of the population would conclude that, if they had to put up with a war again, they would prefer to face the risks of attack by bacteria rather than bombardments by HE.

Fildes evidently took retaliation in kind to be a militarily expedient option, and one to which the majority of the public would acquiesce. Fildes then surpassed Haldane on the equality of chemical warfare with other weapons, and echoed the arguments about bacteriological warfare by Topley, Ledingham and Douglas before the war, in his argument that biological warfare was not merely of the same moral standing as other weapons but should be regarded as a superior and more humane form of killing.[9]

## Medical and Biological Warfare Research

As the war drew to a close, Fildes discussed the future of the Biology Department Porton with his superiors. He envisaged a group reformed under the Royal Army Medical Corps, with close links to the civilian funding agency, the Medical Research Council. Most of the work would be 'pure medical research', with the Services having no authority to initiate lines of enquiry.[10] In Fildes' mind, biological warfare research should be regarded as an adjunct to civil medical research and he wrote to the newly-formed Inter-Services Sub-Committee on Biological Warfare:

> It is perhaps not generally understood that the basic problems of BW are basic problems of medicine. The applications only are different. Applied medicine is primarily concerned with defence, but defence cannot be arranged intelligently without study of offence. Similarly BW is concerned with the exploitation of

9 Fritz Haber, the scientist who invented chemical warfare in its modern form, notably described poison gas as 'a higher form of killing' in 1919. See Harris and Paxman, (1982).

10 TNA, WO188/654. Draft. Post War BW Organisation. P. Fildes. (3 October 1944).

> offence. Thus both the BW workers and medical workers must know how the microbe carries out its offensive activities before the former can exploit them and the latter protect against them.[11]

Fildes believed in the interdependence of biological warfare and medical research, but also knew that his views would not be shared by the Secretary of the Medical Research Council (MRC), Edward (later Sir) Mellanby, and later wrote to the Scientific Advisor at the Ministry of Supply: 'Mellanby, I know, does not want to have anything to do with BW, on what might be called moral grounds, but nevertheless the work is largely medical and the MRC will profit from it'.[12]

Fildes' vision for the future of biological warfare research was not realised, and the department at Porton was reconstituted as the Microbiology Research Department (MRD), answerable to the Ministry of Supply and under the leadership of Fildes' deputy, Dr David Henderson. Many of the staff, who had been seconded from the MRC, returned to civilian research. The MRC, under Mellanby, maintained its distance by reluctantly forming its Biological Warfare Defence committee in 1949 to deal solely with defensive measures. Fildes joined the ranks of scientists returning to civilian research after the war, but actively maintained an input into biological warfare as a scientific advisor.

## Publicity, Recruitment and Secrecy

Although the biological warfare research programme enjoyed a high priority in defence policy and a substantial injection of funds immediately after the Second World War, work at the MRD was dogged by recruitment problems. Scientists were reluctant to work on germ warfare during peacetime, although official discussions of the recruitment crisis tackled the moral dimension of this issue obliquely. For the scientific advisors, attracting new researchers depended on the Government promoting a positive image of biological warfare in the face of adverse media attention.

Problems came to the fore in April 1949, when the Chiefs of Staff Biological Warfare Sub-Committee held a protracted discussion on an anonymous *Lancet* article that denounced all scientific involvement in biological warfare research. After complaining about the ineffectiveness of international declarations to prevent scientists from engaging in bacterial warfare research, the author concluded that: 'we have reached a situation in which men of the highest character are prepared, from a sense of duty, to contemplate what they would normally regard as crimes

11 TNA, WO188/654.BIO/5198. Post War Work on BW. P. Fildes (22 August 1944).

12 TNA, WO188/655. Fildes to Prof. JE Lennard Jones (Ministry of Supply) (13 November 1945).

of enormous magnitude. And this is precisely the state of mind we have so often condemned in our enemies'.[13]

Owen Wansbrough-Jones, a chemist and the Scientific Advisor to the Army Council, spelt out his disdain for the article and its possible long-term consequences to the Biological Warfare Sub-Committee. In his attack, Wansbrough-Jones drew attention to the defensive aspects of the research and, like his predecessors, added that this form of warfare was probably more humane than its counterparts:

> this article typified the need to educate scientific opinion regarding the humanitarian aspects of BW research and the all-important need to provide adequate defences against this form of warfare... the article also ignored the possible development of a highly civilized technique using non-lethal BW agents... If such articles were permitted to pass unanswered they would result in the building up of a mass of opinion in scientific and medical circles hostile to BW which would in due course affect public opinion and be in turn reflected in government policy.[14]

By way of response, the Biological Warfare Sub-committee members suggested that action to 'enlighten scientists should take a subtle form', such as informal talks with the editors of scientific journals or senior members of the British Medical Association. They noted that similar steps had already been taken in the USA and Canada. At a meeting soon afterwards, the members expressed regret that the *Lancet* article had been 'discourteous to those, who through a sense of duty, were engaged in BW research'.[15] They also repeated their suggestion that they should host an informal meeting to 'educate' journal editors, a proposal taken seriously by Sir Henry Tizard, chairman of the Defence Research Policy Committee (DRPC).[16] Tizard also considered whether the news reports might be directly responsible for staff shortages, but dismissed this possibility, informing the Biological Warfare Sub-Committee:

> Since at the present time there is little or no evidence to suggest that the recruitment of staff is being adversely affected by unfriendly criticism in the press, it would be unwise to launch what would be an obvious and perhaps

---

13 Anon. 'Bacterial Warfare', *The Lancet* (5 March 1949), 401–402. The article makes no reference to the 1946–47 Nuremberg Doctors' trial but the language suggests that it may have influenced the writer (see Schmidt 2004).

14 TNA, WO188/663. BW(49)16. BW Sub-Committee. Discussion With US and Canadian Representatives (28 April 1949).

15 TNA, WO188/663. BW Sub-Committee BW(49) 3rd Meeting. (6 May 1949) .

16 At this time, the DRPC had the responsibility of deciding priorities across all non-nuclear defence research (Agar and Balmer 1998).

> provocative propaganda campaign to convert the heretics of the medical world and the scientific press.[17]

Rather than mount a moral crusade, Tizard recommended that the advisors' personal contacts should 'sow the seed quietly'. He added two further measures: that a press release should be written and held in reserve; and that monitoring of press activity on the topic should be maintained.

Early in 1950, the results of this monitoring were released in a report to the Biological Warfare Sub-Committee. A rough count had been taken of the number of articles published in the daily newspapers and medical journals on biological warfare and an increase in coverage duly noted (see Table 2.1). The authors made no attempt to explain the resurgence of articles in 1949, although it is possible that it could be accounted for by news in December from the Soviet Union that Japanese servicemen were being tried for war crimes involving bacteriological warfare (Guillemin 2008).

**Table 2.1 Press Coverage of Biological Warfare**

| | 1946 | 1947 | 1948 | 1949 |
|---|---|---|---|---|
| Specific Articles on Biological Warfare | 20 | 5 | 8 | 33 |
| Passing References to Biological Warfare | 9 | 8 | 6 | 18 |

*Source*: TNA, WO188/664. BW(50)3 (Final) BW Sub-Committee. Public Announcement on Biological Warfare. Report to the Chiefs of Staff. (14 February 1950).

The monitoring report expressed concern that this coverage would prove detrimental to biological warfare and associated research:

> These articles make it necessary to keep in mind the need to condition public opinion, for articles especially such as that in the *Lancet* may result in the growth of opinion hostile to Biological Warfare, with a consequent adverse effect on the recruitment of staff and perhaps on the collaboration of the public in defensive measures at a later stage.[18]

The author then took up the theme from discussions in the previous year and reported that informal routes were already being used to spread the official message. An increasing number of distinguished scientists were 'being consulted

17 TNA, WO188/663.BW(49)21. BW Sub-Committee. Biological Warfare – Publicity (14 June 1949).

18 TNA, WO188/664. BW(50)3 (Final) BW Sub-Committee. Public Announcement on Biological Warfare. Report to the Chiefs of Staff. (14 February 1950).

in the course of research and... given the true facts'. The report also recommended that appropriate information should be passed on to Civil Defence personnel and urged that some public announcement should be made ready to 'correct misapprehensions and to put the subject in its proper perspective'.

A draft press release was prepared and appended to the report. Its main message was that Britain had found it 'necessary to continue research to explore all aspects of Biological Warfare, and so to develop measures for the protection of the Services, the civil population, and the livestock of the country'.[19] The announcement ended by claiming that the research might not only eliminate biological warfare but also have a 'direct bearing on the prevention of disease'. A revised version was soon put together by the Secretary of the Chiefs of Staff Committee in consultation with Tizard of the DRPC and the official committee of Civil Defence. This was a far shorter announcement, once again to be held in reserve and released if necessary. The press release promised that biological warfare would be eliminated and the main paragraph justifying research now read:

> It is the view of His Majesty's Government that the aggressive nature of this form of warfare has been exaggerated; nevertheless it is their duty to do all in their power to safeguard this country against possible attack of this nature. It has therefore been necessary to develop measures for the protection of the Services, the civil population and the livestock of this country.[20]

The announcement went through a further re-draft and research was now promised to 'minimise the significance of BW' rather than eliminate it. The paragraph dealing with the need to continue research had again been revised:

> It is the view of His Majesty's Government that there is no clear evidence that any form of Biological Warfare hitherto suggested would be of major military significance in a country provided with efficient Health Service. However, with the advance of knowledge this may not always be true and it is, therefore, the duty of the Government to do all in their power to safeguard the country against any possible future attack.[21]

By March 1951 the BW Sub-Committee had resolved to keep both of the revised versions in reserve. The Minister of Defence could then choose between them if

19 TNA, WO188/664. BW(50)3 (Final) BW Sub-Committee. Public Announcement on Biological Warfare. Report to the Chiefs of Staff. (14 February 1950). Appended.

20 TNA, WO188/664. BW(50)8. BW Sub-Committee. Public Announcement of Biological Warfare. (22 March 1950).

21 TNA, WO188/664. BW(50)32. BW Sub-Committee. Public Announcement on Biological Warfare (16 November 1950).

it became necessary 'to correct authoritatively any spate of criticism about BW'.[22] This winnowing of information into press releases will be discussed further in later chapters; it is sufficient here to note how such controlled release of information is an opaque window that reveals as well as conceals.

All of this discussion surrounding publicity, while concerned about the moral tenor of criticism, remained focussed on the practical consequences of bad publicity, especially, though not exclusively, for recruitment of new scientists. These on-going recruitment problems spurred Fildes to write a memorandum in May 1951 where, in passing, he linked the crisis to the moral status of biological warfare. Fildes mentioned how it had been thought originally that recruitment through the personal contacts of the MRD's technical advisory committee, the Biological Research Advisory Board, would be sufficient to attract good scientists and '...cause it to be known that first class work could be done at Porton, that professional careers would not be jeopardised and even that there was no moral obliquity involved in preparing to defend one's country'.[23] Rather than pursue this moral line, however, Fildes blamed the failure of personal approaches on the fact that medical salaries had recently been raised above those for the scientific civil service, making it difficult to recruit medically trained staff.

The debate continued, inter-twining the issues of the moral image of biological warfare with recruitment and publicity. Scientists at the MRD planned to invite a hundred scientists to Porton at the end of 1951 in order to demonstrate their work. Very soon, the planners became embroiled in a discussion over whether or not to invite a cohort of journalists. A press release had been drawn up in anticipation of the visit that depicted the MRD as an unremarkable scientific establishment: 'In its broadest sense our work is planned as would that of any other institute devoting its energies solely to the study of infectious disease… the development of specific defensive techniques against possible use of bacteria in war will prove equally important in the control of naturally occurring disease'.[24]

The press release was premature. When the Joint Intelligence Committee was consulted on the security implications of granting the press access to Porton, its members commented that 'if journalists were in the privileged position of being invited to visit MRD they could be told <u>not</u> to publish anything.'[25] In reply, Wansbrough-Jones commented that this apparent gagging order 'appeared to ignore the object of the Press statement. It was not, as appeared to be thought, to focus attention on the MRD but to enlist the aid of the Press in educating the

---

22 TNA, WO188/665. BW(51)10. BW Sub-Committee. Public Announcement on Biological Warfare (15 March 1951).

23 TNA, WO188/668. AC11427/BRB88.BRAB. Memorandum on the Recruitment of Staff for Research in BW by Sir Paul Fildes (3 May 1951).

24 TNA, WO188/665. BW(51)38. BW Sub-Committee. Draft Press Release on Microbiological Research Department. Porton. (24 November 1951).

25 TNA, WO188/665. BW(51) 5th Meeting. BW Sub-Committee (29 November 1951).

public about BW'.[26] He added that a press release would 'guide' the journalists and not be for quoting verbatim. Finally, he reiterated that 'neither should the object of the visit of the scientists be confused with that of the press. The purpose of the former was to assist recruiting; the latter was to enlighten the public about BW'. The visit to Porton took place in December with no media in attendance. Five days afterwards the BW Sub-committee was informed that the press had made no mention of the event.

In the course of planning this visit, the openness of the meeting became an issue which was mixed in with a more general discussion on publicity. Other anxieties exacerbated the issue, particularly concerns over the security of forthcoming sea trials with pathogenic agents off the coast of Scotland, and high profile allegations that the US military had used biological weapons in the on-going Korean War (see Chapter 3). During these debates various scientists and advisors had called for a relaxation of policy on publicity and for a general press release on biological warfare to be issued. The debate was short-lived. In February 1952 the Cabinet Secretary, Sir Norman Brook, wrote a firm letter advising against the press release to the Prime Minister, Sir Winston Churchill. Brook wrote:

> The scientists engaged in this work suffer from a sense of sin which makes them itch to justify what they are doing. Some months ago they sought authority to release a statement claiming that it is more merciful to kill a man by inducing mortal disease than by blowing him to bits with an explosive. I see no reason to suppose that we will have to justify our 'biological warfare' research. And, if we have to do so, I hope we shall not squirm and cringe in the pretence that it is all 'defensive'.... I suggest that these people might get on with their work and stop bothering about publicity for it.[27]

Hand-scribbled at the bottom of this letter is the curt response 'please let them itch a little longer'. And indeed, a few days later a formal letter was written to the biological warfare advisors to inform them that no draft press statement would be approved.[28] Secrecy remained paramount.

## Ethics and Policy

A final hint about the ethical stance of scientists at Porton is gleaned from a meeting of the Biological Research Advisory Board in April 1952. At this meeting, scientists expressed their concern that the US might use biological weapons in

26 TNA, WO188/665. BW(51) 5th Meeting. BW Sub-Committee (29 November 1951).

27 TNA, PREM [Prime Minister's Office] 11/756. Letter from Sir Norman Brook to Prime Minister (23 February 1952).

28 TNA, PREM 11/756. Letter to Brigadier Eubank (26 February 1952).

the on-going Korean war. UK policy, it was stated, was not to initiate biological warfare, which might leave the country in an awkward situation. Not to concur with the United States might entail exclusion from the research collaboration between the USA, UK and Canada. Fildes argued, however, that if the UK did align with the US in the event of their using biological weapons 'this would entail a reversal of Government policy, which would upset research workers at MRD. He (Sir Paul) would feel bound to retire and others with him'.[29]

The Board concluded this discussion by requesting an assurance on British Government policy 'that BW would in no circumstances be used except in retaliation'.[30] The assurance was provided in a letter from Duncan Sandys, the Minister of Supply, who had referred the matter to the Cabinet. This letter was read at the following meeting of the Board.[31] Whether or not Fildes was serious about his threat, the appeal to Government policy suggests that during peacetime that this no first use stance was also a way of justifying engagement with biological weapons research.

## Three Defences

Although these claims about the status of biological warfare differ in historical context and intent between, say, Fildes' impassioned defence of his own work as killing 'without the distressing preliminaries' and press releases designed to promote a positive image of research, three main lines of argument can be discerned throughout these documents. Namely, that biological warfare research was morally justified because it produced humane weapons; it was no different from medical or other research; and it was being performed for defensive purposes. These arguments are scattered sparsely throughout the open literature and it is difficult to gauge whether they enjoyed wider currency. Clearly, the statements on paper were not invented *de novo* and are repeated, which suggests that the quotes discussed here were not the only instances of these arguments being deployed.

The 'humane weapon' line of argument was articulated in two stages. First, according to its proponents, biological weapons were at least as morally good or bad as any other weapon. Second, this assertion was compounded with an argument that biological weapons were actually superior to other weapons in their moral status. The reasons provided for the second component of the argument varied. For Topley and colleagues it was the 'fighting chance' of recovery from infection that bestowed this virtue on bacteriological warfare. With Fildes, it was avoidance of 'distressing preliminaries' that elevated the fruits of his research above other forms of attack on civilians. Finally, for Wansborough-Jones the possibilities afforded by incapacitating, as opposed to lethal, pathogens suggested

29 TNA, WO188/668.AC11872/BRBM25. BRAB 25th Meeting (14 May 1952).
30 TNA, WO188/668.AC11872/BRBM25. BRAB 25th Meeting (14 May 1952).
31 TNA, WO188/668. AC11950/BRBM26. BRAB 26th Meeting (12 July 1952).

that germ warfare was a 'highly civilized technique'. The rhetoric constructed an index for relative suffering, for example through Fildes' comparison of anthrax with high explosive, and thus played with notions of what an honourable use of force might entail.

A second comparison that scientists made was with other research, particularly medical research. This discourse took several forms. According to Fildes, when he sought to reconstitute the wartime research programme at Porton, the research in bacteriology was of such a fundamental nature that there was no point differentiating it from medical research. Fildes went further and depicted biological warfare research as a necessary counterpart to medical science. In the various press statements that were never released, this defence traded on the notion of spin-off, biological warfare research would also benefit civil medical work. Public health in reverse could be reversed a further time. A final use of this approach appeared in the press release intended for the open day in 1951. The MRD was blandly equated with 'any other institute devoting its energies solely to the study of infectious disease.' In all cases, the pursuit of biological weapons was implicitly taken to be a goal overshadowed by alternative benign applications which, to a large extent, were unintended but nonetheless still anticipated.

A final justification recorded in the literature was that research was defensive and undertaken out of a sense of patriotic duty. This argument implied a reversal of the ranking adopted by many scientists, that the supposed purity and international nature of science could be over-ridden by national security needs. Whenever articulated, this line of argument assumed that other countries would have worse intentions for deploying biological weapons than those adopted in Britain. The category of offensive research was never explicitly defined, but acted as an 'unspoken other' against which the label of defense took shape. So, during the Second World War, Hankey regarded as defensive both the policy of retaliation in kind and research to enable such retaliation. Moreover, that retaliation had to be 'in kind' was not challenged at the time in any of the open records. Within this discourse, even research that 'might appear offensive in aim' was permissible because under the appearance of offense were hidden defensive intentions. Put bluntly, no matter what action was taken, the prior statement of policy ensured that it was defensive by fiat.

War provides a peculiar context for ethics and it is evident that the MRC involved itself in germ warfare research only within this context. After the war, Mellanby withdrew the Council from involvement in biological warfare research and was even reluctant to allow for an advisory committee on defensive measures to be formed under its auspices (Balmer 2001). On the other hand, according to Fildes, in peacetime scientists should have been able to work at the MRD and feel a sense of patriotism by defending their country. This view appears to have been shared by other biological warfare scientists at this time. Certainly, when Fildes discussed developments in relation to Korea, he alluded to a general feeling abroad among researchers at the MRD that their work was justifiable only in so far as it was defensive.

Having discussed these arguments, it should be clear that any depiction of the weapons scientist as completely devoid of a sense of moral responsibility is simplistic. This observation is not a defence of biological warfare research. Rather, it is a call to understand that the biological weapon scientists conducted their work in a different, though not necessarily defensible, moral economy. Their justificatory strategies are evidence that complex processes similar to those among nuclear scientists (Gusterson 1996) were at play in the biological warfare community. Although the arguments employed are not formally philosophical, they have their parallels and counter-arguments in the academic literature. Different characteristics of biological warfare are flagged to construe it as being more or less humane than other weapons. The perversion of medical research by biological warfare matches the argument that there is no difference between the two except in application. Finally, defensive research is invoked as a neatly circumscribed moral category, although it is mirrored by the counter argument that defensive research forms a far less tidy empirical category.

However, there is a contextual difference between these arguments. In the archival sources, any defence of biological warfare research was rarely made in isolation. Instead, arguments were deployed in connection with more pragmatic concerns. They found expression in discussions concerning such matters as recruitment and publicity of biological warfare, whether the public would collaborate in defensive measures, the relationship of UK and US policy, and the survival of the programme after the Second World War. The deployment of these arguments cannot be divorced from the instrumental realities of legitimating, maintaining and reproducing a clandestine biological warfare research community – both at the laboratory bench and in the policy arena. These ethical arguments were therefore worked out as what Donovan (2005) calls an 'everyday epistemology', not in abstraction but as just one element of a community worldview that connected with many material, social and psychological commitments.

This is not to suggest that the moral dimension was a foil for the real goals of the workers and that, deep down, Fildes and other scientists somehow really knew what they were doing, while putting up an ethical smokescreen. A close reading of these arguments, together with their occurrence alongside instrumental goals, suggests a different point. Although the justificatory arguments appear to have been directed primarily at the 'heretics', opponents of biological warfare would have constituted an unlikely audience. The research was shrouded in secrecy, and policy matters were discussed with even greater attention to security. Silence was generally the response to criticism. As we shall see in chapter six it is only in the late 1960s that, accompanied by increasing media and public attention, the Ministry of Defence and staff at Porton began to pay close attention to public relations strategies with respect to the Porton establishments.

Instead, during the period we have discussed, statements about the desirability of biological warfare, whether in committee papers or unreleased press releases, would appear to have been used more within the biological warfare community. As inwardly directed narratives, defences of biological warfare research would have provided a ready resource, a 'tool-kit' of rationales and meanings for those engaged in the work. In other words, these arguments were both an expression and a means of perpetuating the moral economy of the biological warfare community. Their audience was proximate. If this is the case, then Norman Brook, in declaring that these scientists 'suffer from a sense of sin', was invoking a comfortable stereotype and nothing more. Likewise, if we regard scientists engaged in biological warfare research as simply violating universal professional norms, then we may have bought into the same stereotype. It would be less comfortable, but make more sense, to recognise and then dismantle the weapon scientists' engagement with their alternative and secret moral economy.

# Chapter 3
# Making Secrets: Accidents, Experiments and the Production of Knowledge

During the Cold War, an integral part of the British research aimed at developing an anti-personnel biological weapon involved a series of outdoor trials at sea where the scientists exposed animals to clouds of pathogenic (disease-causing) micro-organisms. Not all of these trials ran according to plan and at the culmination of 'Operation Cauldron', a series of trials that took place between May and September 1952, something quite unexpected occurred. On the final day of the trials, a fishing vessel, the *Carella*, strayed into the danger zone around the trial shortly after a test bomb had been detonated. The *Carella* ignored warnings and the ship's crew was exposed to an invisible cloud of germs. The authority's response was to tail the trawler over the next month without notifying the crew, and to listen for a distress call. The Navy also drew up contingency plans in case the crew fell seriously ill. These plans attempted to balance effective treatment of the victims with security considerations. The incident was so sensitive that almost all records of its occurrence were burnt; the single remaining file on the incident remained locked within the Ministry of Defence for fifty years.

Apart from a brief article in the British newspaper, *The Observer*, in 1985, which generated some fruitless questioning from the local Member of Parliament (MP) and a few briefer articles in the local press, together with a short description in the context of a discussion of the complete series of Scottish biological warfare trials, the events surrounding the *Carella* have not been recounted and analysed in detail (Anon 1985, Leigh and Lashmar 1985, Taylor 1985, Willis 2003, Wilson 1985). This chapter explores the decision-making process during the incident and the wider implications for our understanding of science and secrecy.

The account draws attention not simply to the utmost secrecy surrounding these events, where secrecy became a guiding principle that overrode other considerations, but equally to demonstrate how secrecy operates and what secrecy allows. I argue that two processes are apparent for maintaining secrecy. First, actors treat the secret as a self-contained thing, a singular entity. Second, those who hold the secret attribute a tremendous amount of knowledge and insight to those people outside of the culture of secrecy. A different strand of my analysis emphasises that 'what secrecy allows' should be understood not only in a restrictive sense, but also as productive of effects. In this respect, the existence of secrecy permits familiar categories to blur, in particular those of accident/experiment and patron/scientist. As categories blur, secrecy allows for a particular form of knowledge production that, in turn, engenders further secrecy. In short, they are co-produced.

Because this case involved exposure of the public to biological warfare agents, a related question arises about what activities and interventions take place, and are even legitimated, under the protection of secrecy. With respect to risks, this question is frequently asked within empirical studies of human experimentation (Schmidt and Frewer 2007, Goodman, McElligott and Marks 2003, Oakley 2000, Jones 1993, Moreno 2001). In these studies, discussion often focuses on the concept and limits of informed consent, with infamous cases such as the Tuskegee syphilis experiment or Cold War radiation experiments often used as the exemplars of what medical experimenters find permissible when secrets are kept from 'guinea-pigs'.[1] Significantly, these studies note that the term 'guinea-pig' encapsulates the dehumanisation of the experimental subjects permitted by these secretive conditions. And, despite an emerging cadre of professional volunteers for current biomedical research experiments appropriating the term 'guinea-pig' (Weinstein 2001), it still retains strong pejorative connotations (Moreno 2001: 276–81).

Both the pejorative and positive appropriations of the 'guinea-pig' label underline the variety of ways that people become experimental subjects. Gray, based on interview data with clinical trial participants, has developed a typology that, while somewhat static and ideal-typical, begins to illustrate this variety (Gray 1975). His spectrum consists of: unaware subjects who have no knowledge of being part of an experiment; unwilling subjects forced to participate; benefiting subjects who gain from their participation; indifferent subjects who follow the doctor's instructions; and committed subjects who stated that altruism was their motivation to enter the research. At first glance, the fishermen in this case study would seem to have entered the experiment as unaware guinea-pigs.

Yet what might be worse than being a guinea-pig, is being a 'fly in the ointment'. In other words, you could be someone who simply gets in the way of an experiment. As such, it is important to distinguish between participants in the 'unaware' category. There are those who researchers intend to be part of the experiment, such as the patients in the Tuskegee experiments, but also those who become part on an on-going monitoring experiment by being 'in the wrong place at the wrong time', as was the case of the longitudinal studies of the health of Pacific Marshall Islanders exposed to fallout from nuclear weapons tests (Crease 2003). Furthermore, this division should be seen as permeable. For example,

1 The notorious Tuskegee experiments ran between 1932 and 1972 (Jones 1993). Over 400 black people in Alabama were observed by medics as their syphilis progressed. They were told they had 'bad blood' but not told nature of their condition, and were given tests but not given penicillin when this became available as a treatment. Various human experiments took place with radioactive materials during the Cold War. For example, following safety concerns during the Manhattan Project, after the Second World War seventeen patients, many with forms of cancer, were injected with doses of plutonium and samples taken from them (blood, bone, excrement, urine) to see what happened to the plutonium (Moreno 2001).

when the British Ministry of Defence was accused of using soldiers as 'guinea-pigs' by making them run and crawl through atomic bomb sites in the Southern Australian desert, their response was that they were not testing people, only their clothing (Mitchell 2003). In the case study in this chapter, which can be located in this 'participation by getting in the way' category, I will argue that those who are accidentally or deliberately exposed, but are not the objects of experiment, challenge not only the boundaries between accident and experiment (Levidow 1990), but, significantly, between the patron and experimenter.

This challenge to the boundaries between experiment and accident is readily linked with secrecy. I stated earlier that part of the aim of this book is to explore what secrecy allows. While this can be readily construed in a negative sense, as a prohibition on action, it is not the only sense in which we can understand what secrecy permits. Secrecy, as I will argue, may also permit certain effects in a more productive sense, by blurring categories and acting as a guide or compass to orient behaviour. By doing so, secrecy may allow and sustain particular forms of knowledge production.

A helpful way of theorising this link is to borrow from recent developments in science studies and regard secrecy and science as *co-produced* or *co-constructed* (Jasanoff 2003, Irwin 2001). The co-production framework refuses to take the categories of the 'technical' and the 'social' as *primum mobile*, prime or unmoved movers, that can be called on so as to explain either science wholly in terms of social factors or science purely in terms of more science. Rather, as Shapin and Schaffer have argued, 'solutions to the problem of knowledge are solutions to the problem of social order' (Shapin and Schaffer 1985: 332). They work simultaneously. To take a recent empirical example of co-production and its discontents, Reardon has demonstrated how the Human Genome Diversity Project, the plan to catalogue our world-wide genetic variation, encountered strong and continued opposition by addressing issues, such as colonisation, intellectual property and the origins of human diversity, as either exclusively social or exclusively scientific (Reardon 2001). Yet human genetic diversity, she argues, cannot be neatly sorted in this way when, for instance, questions about what groups to sample, in order to answer questions about human origins, are entangled with questions over what groups to seek consent from, in order to resolve consent or intellectual property issues. Specifically, Reardon argues that two different sampling techniques were available: either take genetic material from individuals at points from a grid or sample from existing groups. The first method implies seeking out individuals for their consent; the second implies approaching spokespeople or community leaders for group consent. Each 'technical' question is simultaneously an 'ethical' question. Moreover, scientists' and lawyers' answers to these questions did not neatly map on to the answers proposed by indigenous peoples. The project organisers attempts to settle first one, and then the other set of issues did not work and, Reardon claims, this failure was because these scientific and social issues were inseparable as they continually shaped and redefined each other. Put in Jasanoff's terms, 'the realities of human experience emerge as the joint achievements of scientific, technical and social enterprise: science and society, in a word, are co-produced, each underwriting each other's existence' (Jasanoff 2003: 17).

Returning to the relationship between science and secrecy, whereas the question of what secrecy permits is usually, and quite rightly answered negatively in terms of loss of accountability, secrets can equally be seen to permit the elision of ordinary categories. I will argue that secrecy ensured that the accident with the trawler became a monitoring experiment. Social and natural order were co-produced as secrecy entailed a particular set of arrangements for producing knowledge about biological weapons, and this arrangement also engendered further secrecy. Beyond this, the academic literature has less to say and so we turn back to the case study in hand to see what happens in such circumstances.

## Secret Preparation for Operation Cauldron

An anti-personnel biological bomb remained the central goal of the early Cold War biological warfare research programme. Outdoor trials were essential for testing agents and munitions and the 1952 biological warfare trial codenamed 'Operation Cauldron' had been preceded by a trial in the Caribbean, off Antigua, between December 1948 and February 1949. 'Operation Harness', as this first post-war trial was codenamed, had involved around two years of planning, elaborate secrecy, some 450 personnel, two ships and numerous experimental animals. During Harness, scientists systematically exposed sheep, monkeys and guinea pigs to a range of pathogenic organisms including bacteria responsible for anthrax, brucellosis and tularaemia (Balmer 2001, Hammond and Carter 2002).

Even before Operation Harness finished, the scientists at Porton had laid plans for a further series of trials, Operation Cauldron. Two years and several setbacks later, by late 1951, a trial site off the Hebridian island of Lewis in north-west Scotland had been located. In a similar vein to Harness, these trials were to involve scientists and navy personnel releasing pathogens from test munitions in the vicinity of the test animals harnessed on a floating pontoon. A naval vessel, HMS *Ben Lomond* had been allocated as the main ship that would be used during the trial. Although there was never any question that Cauldron would be undertaken in utmost secrecy, international events during 1951 meant that the trials would occur while biological warfare was an especially sensitive topic.

On 8 May 1951, a little over a year into the Korean war, the North Koreans made their first of several allegations that the United States was employing biological weapons in Korea (Farrar-Hockley 1995:279–280, Leitenberg 1998, Moon 1992). Further denouncements followed throughout 1952. While the claims were generally dismissed in Britain, they had been afforded some credence by Dr Joseph Needham, a Fellow of the Royal Society. Needham was the British delegate to an International Scientific Commission tasked by the, predominantly communist, World Peace Council to investigate the biological warfare charges (Buchanan 2001). With Needham's adverse publicity, British troops committed to Korea, a potentially hostile reaction to the charges across Asia, and moreover, behind closed doors, with strong collaborative ties between the British and US biological warfare programmes, any sudden revelations about domestic research into germ warfare at this time would have been especially unwelcome for the Government (Carter and Pearson 1996, MacDonald 1990).

Before Operation Cauldron could proceed, the Admiralty and Scottish Fisheries Office became involved in extensive negotiations, intended to ensure secrecy with minimal disruption to the fishing trade in the area. These negotiations finished in March 1952, when the Admiralty was able to confirm the various conditions under which the trials would be conducted. Trials were to occur between 8am and 8pm. Ships would patrol the limits of a danger area, set at 5 miles from the test site. Warnings had to be provided both in advance and at the time of trials, and the Admiralty was instructed to give notice to Fisheries officers concerning the timing of the trials. In this respect, the Admiralty was instructed that 'no start with trials [are] to be made so long as any fishing boats are within the danger area' and that they were to 'accept responsibility for any accident arising in the trial area and [which was] directly attributable to Admiralty operation of the range'.[2] Local fishermen received notice of the trials from the Admiralty in the form of a chart of the danger area and a warning that 'special trials' would be taking place from May to September. The notice informed fishermen that they were to steer clear from the five mile radius of the danger area when a red flag was flying. Although there was no hint whatsoever that the trials involved biological agents, the commanding officer of *Ben Lomond* had still recommended that the warning was 'couched in as strong language as possible in conformity with the requirements for security'.[3]

The warnings for fishermen were not the only sources of information kept to a minimum. Discussions took place back in Whitehall concerning any wider publicity that the trials might attract. Winston Churchill, in his capacity as Minister of Defence, had already been advised in February 1952 that the previous Labour Government had held to the policy that 'no official statement should be made on BW unless alarming articles started appearing in the Press'.[4] However, by June one such alarming article was on the horizon, with the *Daily Express* threatening to publish information about the month old trials. This led civil servants to suggest that a press release should now be issued. Indeed, a draft for just such an occasion had already been prepared for the Chiefs of Staff's Biological Warfare Sub-Committee in January. The draft emphasised defence, stating that biological warfare research had been taking place in Britain since the Second World War and:

---

2 TNA, ADM[iralty] 1/26857. Memo from Department of Health for Scotland (25 March 1952).

3 TNA, ADM 1/25255. From Commanding Officer, HMS Ben Lomond to Flag Officer, Scotland. Cauldron – Operational Trials Area – Warning to Fishermen (14 May 1952).

4 TNA, PREM 11/756. R.W. Ewbank to Minister of Defence. Publicity for Biological Warfare Trials (21 February 1952).

> in order that effective means of defence may be developed, every possibility must be studied, not only in the laboratory but in the field... for safety these experiments will be carried out at sea. Only by such means can the risk from biological warfare attack be adequately assessed and specific defence measures perfected.[5]

Even with the stress on defense, the Prime Minister, Winston Churchill, firmly opposed this or any kind of press release. On being informed by Lord Alexander, who had taken over as Minister of Defence in March 1952, of the imminent *Daily Express* article and the existence of the draft statement, Churchill scribbled at the bottom of Alexander's memo 'I don't like the statement at all'.[6] Churchill did not let the matter rest and, that same day, wrote a memorandum to Alexander and Sir Ian Jacob, Chief Staff Officer to the Minister of Defence, suggesting that:

> nothing should be done until there has been a publication, through leakage, in the Press. After that no doubt there will be a Question in Parliament and it is then that the answer can be considered. Any statement made by us would naturally base itself on the fact that the late Government had sanctioned the proceedings, and that the measures were purely defensive. All details should be avoided...[7]

Churchill's reluctance notwithstanding, his advisors continued to proffer him draft press statements for consideration. The first, which is worth quoting at length, contained a substantial amount of information:

> The Communists have been waging, and are continuing to wage, a campaign about the use of bacteriological warfare in Korea. The accusations they have made are entirely false. Nevertheless, in our experience of dictators, we have found that when a totalitarian Government has it in mind to perform some outrageous act they usually think it good tactics to accuse the other side of the very crime they propose to commit. Furthermore, it is noteworthy that the Soviet Government made a specific reservation to the 1925 Geneva Protocol (which forbade biological warfare) and said they would cease to be bound "vis a vis any enemy state whose armed forces or whose allies did not respect the provisions of the protocol". The Soviet Government might falsely claim that the U.K. was allied to a power that had used bacteriological warfare. For these reasons HMG have thought it wise to follow the example of the

---

5 TNA, WO 188/668. BW(52)1 BW Subcommittee. Field Trials with BW Agents (29 January 1952).

6 TNA, PREM 11/756. Lord Alexander to Prime Minister (25 June 1952).

7 TNA, PREM 11/756. From Churchill to Minister of Defence, General Jacob. (25 June 1952).

> late Government in conducting research into methods of defence against the possibility of aggression by such means.[8]

Such candour did not stand a great chance of reaching the press. Even a more modest draft press statement fared little better. Lord Alexander proposed that in case of a press leak something should be held in reserve that read:

> The possibility that bacteria may be used in a future war cannot be overlooked. Researches are being conducted so that we may be ready to meet any situation which may arise. The experiments now taking place form part of these researches. Effective means of defence against bacteriological warfare must be developed, not only in the laboratory, but in the field.[9]

Churchill stripped this down to a stark and uninformative statement, only to be used in response to an unwelcome question in the Commons:

> The possibility that bacteria may be used in a future war cannot be overlooked. The researches begun by the late Government are being pursued so that defensive measures may be taken. The experiments now taking place form part of these researches.[10]

It is important to avoid an anachronistic reading of these statements, Churchill and colleagues were not simply being disingenuous by labelling the experiments as defensive, despite their overall objective of developing a workable biological weapon. His understanding, which dated back to the Second World War, was that defensive research could equate with the 'use in retaliation only' policy that had prevailed since the inception of the BW research programme (Balmer 2001). This said, the labelling of these experiments as defensive highlights the blurred boundary between offensive and defensive research (for further discussion see Piller and Yamamoto 1988, Strauss and King 1986, Wright 1990).

By the time Churchill produced his desiccated publicity statement, Operation Cauldron had been underway for a month. Scientists from Porton had rigged up a large floating pontoon where they placed monkeys and guinea-pigs in cages (Hammond and Carter 2002). A five mile exclusion area around the pontoon was patrolled by the 'clear range' vessels. Closer to the pontoon, the commanding officer of HMS *Ben Lomond* would give the order to release potential biological agents at a distance of 25 yards from the animals, the bacterial cloud blowing over them and then out to sea. *Ben Lomond* was then used as the base for collecting and carrying out autopsies on the animals. Press attention to the trials turned out to be scant, with a few articles appearing but attracting no further attention (Willis

---

8 TNA, PREM 11/756. JRC to Prime Minister (26 June 1952). Emphasis in original.

9 TNA, PREM 11/756. Lord Alexander to Prime Minister (26 June 1952).

10 TNA, PREM 11/756. Churchill to Minister of Defence (26 June 1952).

2003). Even the local paper had little to add on the trials, 'nothing was said about them in the *Stornoway Gazette*, which only referred to the *Ben Lomond* on the rare occasions when a football team from the ship could be landed at Stornoway' (Willis 2003: 288). And so the trials ran throughout the summer and into September.

## The 'Carella' Incident

On the last day of the Cauldron series of trials, a secret signal was sent from the captain of the *Ben Lomond* to the Admiralty. The signal contained a cryptic but worrying message:

> During Cauldron trials using agent L at 1900 15th September, steam trawler CARELLA Number H4 of Hull bound Fleetwood from Iceland(C) disregarded signals and crossed danger area after release of agent. Vessel passed two miles to leeward position of pontoon sixteen minutes after time of release. Wind speed six knots. Consider vessel may have passed through toxic cloud.[11]

According to later reports, the fishing trawler, with 18 men on board, had approached from the north and been sighted just a minute before seven o'clock by one of the boats patrolling the danger area.[12] Two minutes later, the crew of the *Ben Lomond* sighted the trawler some two miles distant and it was considered 'not likely to be in danger for another quarter of an hour'.[13] The commanding officer on the *Ben Lomond* judged that he could use signals to divert the trawler. So, with all the trial ships flying red warning flags and a searchlight repeatedly signalling 'You are standing into danger', the captain proceeded with the trial. The biological bomb was detonated at two minutes past seven as the trawler, ignoring or not having seen the signals, continued within two miles downwind of the pontoon, sixteen minutes after the bomb had been detonated, and 'possibly through the fringe of the toxic cloud'.[14] The agent used in that particular explosion was codenamed L, the top secret name used for *Pasteurella pestis*, the bacteria responsible for plague.[15]

Indeed, the identity of L was such a closely guarded secret that the significance of the signal sent from the *Ben Lomond* was initially not recognised. The head of the Military Branch at the Admiralty later explained that: 'The signal which was Secret had no priority marking whatever and the only word which could have

---

11 TNA, ADM 1/26857. Signal From BEN LOMOND to Admiralty. (16 September 1952).

12 TNA, ADM 1/26857. The Cauldron Incident. Admiralty. (n.d. 1952).

13 TNA, ADM 1/26857. The Cauldron Incident. Admiralty. (n.d. 1952).

14 TNA, ADM 1/26857 The Cauldron Incident. Admiralty. (n.d. 1952); ADM 1/26857. Events of 15th September (n.d. 1952).

15 In 1970, *Pasteurella pestis* was renamed *Yersinia pestis*.

alerted anyone to its importance was "toxic" which came at the very end'.[16] The coded signal had been received in the Wireless Room at the Admiralty at 0220 hours and then had to be deciphered and typed up. It was not until noon that anyone who knew about the trials actually saw the decoded signal, and even then it was not until a phone call had been made to scientists at Porton Down that Agent L was identified as the plague bacterium.

The return signal, sent to the *Ben Lomond* on the same day, contained a short and unruffled response to the incident: 'C.S. M.R.D. [Chief Scientist. Microbiological Research Department] Porton considers danger negligible. Admiralty concurs. No further action required by you or F.O. [Flag Officer] Scotland'.[17] But this was by no means the end of the matter. To begin with, the safety of the trawler had been presumed on a reported wind speed of six knots. A memo written a few days after the incident noted that the original estimate referred only to the situation at the pontoon, whereas over the whole two miles between the pontoon and trawler the wind speed was more likely to have been anything from five to nine knots. With the estimate revised, the memo admitted that 'there now is a possibility that [the] vessel passed through the edge of the toxic cloud'.[18] To make matters worse, the estimated two mile distance of the *Carella* from the pontoon was now qualified with a maximum margin of error of 400 yards.

### *Whitehall Responds*

Information about the incident was duly passed to the Minister of Supply, Duncan Sandys. He held a meeting on the evening of the 16th September with representatives from the Admiralty, Ministry of Supply and Ministry of Health. The Ministry of Supply was responsible for Porton Down and Dr David Henderson, superintendent of the MRD, was present at the emergency meeting. At the meeting, the discussion focussed on the potential risk to the *Carella* crew and the precautions that might be taken against plague breaking out.

The following day, Sandys conveyed the conclusions of this meeting to Chancellor of the Exchequer, Richard Butler, standing in for the Prime Minister.[19] In his report, Sandys noted that the trawler could not have passed through the toxic cloud *if* the estimates of its course and the wind speed were correct. On the other hand, the possibilities that the distance of the trawler from the pontoon had been overestimated, or the wind-speed underestimated, meant that the trawler

16 TNA, ADM 1/26857. Head of Military Branch to Deputy Secretary (Admiralty) (17 September 1952).

17 TNA, ADM 1/26857. Signal From Admiralty to Ben Lomond (16 September 1952).

18 TNA, ADM 1/26857. Events of 15th September (n.d. 1952).

19 Churchill, who had been suffering ill health for some time, spent the whole of September on vacation in the south of France at the villa of his friend, Lord Beaverbrook, (Young 1996: 90).

might have passed 'through the edge of the toxic cloud for a short time'.[20] Sandys added that laboratory data indicated that the plague bacilli would only survive for a few minutes once exposed to the air, whereas the trawler had come closest to the pontoon after sixteen minutes. He therefore concluded:

> Assuming however that the trawler did in fact pass through the edge of the toxic cloud, which is unlikely, and assuming that all the bacteria were alive when they reached the ship, which is highly improbable, the low density of the bacteria at that distance (about 2 miles from the point of release) would make the chance of infection extremely low indeed.[21]

Sandys then moved on to inform Butler of possible precautions that could be taken against the plague. First would be to provide the trawler crew with a course of streptomycin injections for several weeks and to keep them under medical surveillance. Second, the trawler and crew would need to be decontaminated by killing any rats on the ship and any fleas and other insects on the fishermen. Both courses of action would entail some compromise of the elaborate secrecy surrounding the trials. Sandys was well aware of this problem, adding in his memorandum that it would be impossible to conceal that these precautions were being taken against infection. More specifically, Sandys argued that if the authorities were to eradicate the rats on the *Carella*, this would suggest that they were concerned about plague. So, he warned, 'in these circumstances it would not be possible to conceal the reasons for this action or devise any effective cover plan'.[22] The Minister then added that because the toxic cloud would have been invisible, it was safe to assume that the crew of the *Carella* were unaware that any trials were in progress when it entered the danger area. In the light of all of this information, Sandys concluded by allowing security considerations to prevail and, quite literally bracketing out his involvement, suggested a course of inaction: 'My own opinion, though I have no responsibility in the matter (beyond that of having manufactured and released the bacteria) is that the incident should be ignored and that nothing should be said or done about it'.[23]

It is not entirely clear whether or not Butler had received this memo from Sandys when he held a meeting on the incident at 2.30pm on 17th September. When the meeting was arranged, the Admiralty had been operating under the understanding that the *Carella* would return home to Fleetwood, Lancashire, late

20 TNA, ADM 1/26857. Memo Duncan Sandys to Chancellor of the Exchequer (17 September 1952).

21 TNA, ADM 1/26857. Memo Duncan Sandys to Chancellor of the Exchequer (17 September 1952).

22 TNA, ADM 1/26857. Memo Duncan Sandys to Chancellor of the Exchequer (17 September 1952).

23 TNA, ADM 1/26857. Memo Duncan Sandys to Chancellor of the Exchequer (17 September 1952).

that evening or early the next day. It was decided that although no immediate action was necessary, some preparations should be made in case infection occurred either in port or on the trawler's next outing. While the apparatus of the Cauldron trials was being dismantled, the decision to monitor the *Carella* without intervening extended the ambit of experiment in an unforeseen direction.

Once again for the authorities, maintaining secrecy was of paramount importance. At the meeting arranged by Butler, Ministry of Health officials were instructed to 'warn local representatives confidentially, without telling them the reason why'.[24] In addition, they were to arrange for supplies of streptomycin to be held in reserve at Fleetwood, also in Iceland and at unspecified points along the Scottish coast. It fell on the Admiralty to arrange for a ship, equipped with a medical officer and streptomycin, to 'be at seas in the vicinity within short streaming distance of CARELLA during her next fishing trip'. Finally, the representatives at Butler's meeting agreed that both Ministries would obtain, with utmost discretion, the names and addresses of the crew.

In fact, the *Carella* had already arrived in Fleetwood ahead of schedule. While this caused some consternation at Ministerial level, on the grounds that no action could now be taken before the crew disembarked, the Admiralty appealed to secrecy in order to argue that this knowledge would have made no difference. The Head of the Military Branch later wrote:

> While I regret that my colleagues were thus not informed of all the possibilities, I find it hard to believe that this made any practical difference to their decision. No overt action could have been taken on or before the arrival of the trawler at Fleetwood without publicity and the assessment of the situation that was expected was that the chances of infection were so remote as not to justify the positive disadvantages of publicity about these trials in general and this incident in particular.[25]

Following Butler's meeting, the Chancellor wrote an additional memorandum to Commander Allan Noble, a conservative MP and Parliamentary Secretary at the Admiralty, emphasising that he was relying on the Admiralty to deal effectively with any outbreak of plague that might occur. Butler stressed that any such arrangements were to be made with due reference to security:

> I will not go into detail as to how this could be organised, but you will presumably have regard to the need for secrecy, and will rely on the ship's radio. You will,

---

24 TNA, ADM 1/26857. Initialled A.N. [Commander Allan Noble] to First Sea Lord (18 September 1952).

25 TNA, ADM 1/26857. Minute from Head of Military Branch (23 September 1952).

> presumably, also remain in touch with the health authorities with a view to having the necessary drugs in the case of trouble arising.[26]

The next day, 18th September, Noble replied with a detailed series of arrangements for tailing the *Carella* and monitoring for signs of plague. More drugs were to be sent to the British ambassador in Iceland, although this instruction was overridden the next day at the behest of the Foreign Office. Two vessels were now to become involved in the pursuit of the trawler, which was due to leave for Iceland on Saturday 20th September. A destroyer, the HMS *Zambesi*, would follow the trawler to the fishing grounds, then a Fishery Protection Vessel, *Truelove*, would receive the medical officer from the *Zambesi* and remain close to the fishing grounds. In both cases, wrote Noble: 'these ships will be given comprehensive instructions as to secrecy and will in no way shadow or escort the trawler, but rely on her wireless signals. Arrangements have also been made for a listening watch to be kept on the trawler wave from shore'.[27]

Finally, Admiralty staff had been instructed to approach the owners of the *Carella* at the Dinas Steam Trawler Company. They were told to obtain the names and addresses of the trawler crew, plus details of any changes of personnel that might have occurred. Noble informed Butler that the owners would be told a cover story, 'that the trawler was in a forbidden area and that the men may have seen secret equipment which it was particularly desired that they should not; this is, in fact, true'.[28]

*More Orders, More Secrets*

It was now time to issue special top secret instructions to the Medical Officer and the Commanding Officer of the *Zambesi*. In these orders, the Medical Officer was told that the crew of the *Carella* may have been contaminated by a bacterial bomb containing *Pasteurella pestis* and that he was to be prepared to make a diagnosis of pneumonic plague. The officer was also given instructions for treating any victims with streptomycin, syringes and needles that had been supplied with his orders. At this point, the officer was also informed that there was a maximum incubation period of 21 days for the disease, so that the danger period from the disease extended, with military precision, until exactly 19.00 hours on Monday 6th October. Whatever transpired in this period, under no circumstances were the crew to be told of the danger that they were in or the details of any illnesses:

> The fact that this exposure may have occurred is not at present known to the crew and is on no account to be revealed to them in any casual contact. Should

26 TNA, ADM 1/26857. R. Butler, Chancellor of the Exchequer, Treasury Chambers to Commander Allan H.P. Noble (17 September 1952).

27 TNA, ADM 1/26857. Letter Noble to R.A. Butler M.P. (18 September 1952).

28 TNA, ADM 1/26857. Letter Noble to R.A. Butler M.P. (18 September 1952).

> you be called to a case which you suspect, ought to [sic] diagnose as Plague, it should be referred to as pneumonia in your dealings with the crew.[29]

Finally, as if to underline the clandestine nature of the whole operation, the officer's instructions were 'to be destroyed by fire' the moment the official danger period expired.[30]

The commanding officer of the *Zambesi* was given even less detail. The Admiralty informed him that the crew of the *Carella* 'may have been in contact with virulent germs from a service source. Crew are not aware that they may have been exposed to infection and this is not to be revealed to them in any casual encounter unless specifically ordered to do so'.[31] The *Zambesi* was ordered quite specifically not to appear as to be visibly shadowing or accompanying the *Carella*. Instead, they were told to constantly monitor the trawler's radio frequency in case the crew reported sickness or the need for medical assistance. At this point, the *Zambesi* would intervene, provide medical aid and transmit an urgent and especially encrypted signal conveying the news back to the Admiralty. Otherwise, on reaching Iceland, the medical officer would be transferred to the *Truelove*, which would take over the role of the *Zambesi*. At this point, however, the commanding officer of *Truelove* had been told nothing about the situation.[32]

Even as these instructions were being transmitted from the Admiralty, more details about the accident emerged. A report about the exposure had been compiled by Captain Welby-Everard of HMS *Ben Lomond*. In the report, he noted that the agent had been released in two separate tests at 18.09 and 18.39 hours. There had then been pressure to proceed rapidly with a third test because favourable conditions were unlikely to last for long at that time of day. In addition, the weather forecast for the following day predicted unfavourable conditions for the trials.

Once the *Carella* had been spotted, Welby-Everard judged that he would be able to stop the itinerant trawler. Earlier in the course of Operation Cauldron, he admitted, there had been a similar situation and he had successfully stopped and diverted the stray vessel. So, the Captain ordered the plague bacteria to be released, and danger signals were flashed at the trawler. Ten minutes later, with no response from the *Carella*, 'it was decided that it would be better to allow her to continue than perhaps to stop her in a position to the leeward of the pontoon'.[33]

---

29 TNA, ADM 1/26857. Special Instructions to the Medical Officer (18 September 1952).

30 TNA, ADM 1/26857. Special Instructions to the Medical Officer (18 September 1952).

31 TNA, ADM 1/26857. Top Secret Signal. Admiralty to ZAMBESI (19 September 1952).

32 TNA, ADM 1/26857. G. Barnard to Parliamentary Secretary. Trawler 'Carella'. Situation at 1100 Saturday 20th September (20 September 1952).

33 TNA, ADM 1/26857. Report on Incident with Trawler Carella During Operation Cauldron. Captain PHE Welby-Everard. (19 September 1952).

As the *Carella* passed the *Ben Lomond*, no one was seen on the trawler's deck. Eventually, the *Carella* was intercepted by one of the patrol ships, the *Hengist*, and the surprised fishermen explained to the crew of the *Hengist* that they had called *Ben Lomond* but received no response. According to Welby-Everard, there was 'no doubt that she did not call *Ben Lomond* until well into the area as she was under continual observation through binoculars from *Ben Lomond*'s Bridge for at least 15 minutes from first sighting'.[34]

The Admiralty had also received another report, from the scientific trials officer, John Morton. His report noted that previous tests in the Cauldron series, using plague to try and infect monkeys, had resulted in three monkeys being infected at doses of between 0.002 and 0.005 mg.min./$m^3$. On the other hand, five monkeys that had been given doses of between 0.001 and 0.002 mg.min./$m^3$ had not developed plague. Morton also calculated the maximum dose at two miles from the pontoon, assuming that the bacteria did not die during downwind travel, to be about 0.0000002 mg.min./$m^3$. Of course, he added, the plague bacilli would be killed during downwind travel and this would lower the dose to which the *Carella* had been exposed. Morton concluded that it was most unlikely that any organisms could have been detected near the *Carella*, and added, with no hint of irony, that: 'even with the highest possible dosage at that distance it would have been virtually impossible to infect experimental animals'.[35]

### *Tailing the Trawler*

On Monday 22 September, another top secret signal was sent from the Admiralty to the *Zambesi*, this time containing instructions that they were to pass by hand to the *Truelove.* In addition to the information supplied to the commanding and medical officers of the *Zambesi*, the same officers of the *Truelove* were told that the danger period would run until the 6th October and that this danger was diminishing by the day. Besides the instruction to provide any required medical aid to the crew of the *Carella*, the *Truelove* was also ordered to prevent the trawler from entering any Icelandic or other foreign port. Instead, if disease did break out, the *Truelove* was to escort the stricken trawler back to a British port. The Admiralty's priorities were spelt out clearly in the signal. *Truelove*'s commanding officer was informed bluntly that, by preventing *Carella* from landing anywhere except back at home,

34 TNA, ADM 1/26857. Report on Incident with Trawler Carella During Operation Cauldron. Captain PHE Welby-Everard. (19 September 1952).

35 TNA, ADM 1/26867. Report by Scientific Trials Officer. J.D. Morton (19 September 1952). Poor recovery rates for *P. Pestis* were eventually attributed to high humidity; Paul Fildes, the former head of the biological warfare research programme, later declared the organisms a 'failure' in operation Cauldron (Balmer, 2001: 118).

'grave political requirements are more important than what may seem best from a strictly medical point of view for individuals'.[36]

As the oblivious crew of the trawler continued north, back in Whitehall an attempt was made to apportion blame for their accidental exposure and potential contamination. According to the Deputy Chief of the Naval Staff, the commanding officer of the *Ben Lomond* had made a 'serious error of judgement' and that his 'keen-ness outran his judgement'.[37] This censure was qualified somewhat as the Deputy Chief of the Naval Staff added that this had been a dangerous series of trials involving intricate safety precautions.[38] Moreover, the scientists and naval personnel had been kept safe throughout the operation. With secrecy once again taking precedence, he recommended that holding a board of inquiry would 'be unlikely to find out anything new and would be attended by great security difficulties'.[39]

At a meeting on Wednesday 24th September, the Lords of the Admiralty approved a letter of reprimand to Welby-Everard, commanding officer of the *Ben Lomond*.[40] The letter informed Welby-Everard that he was guilty of an error of judgement; he was also chastised for not highlighting the urgency of the situation in his initial signal sent to the Admiralty after the incident. The letter was held back until the 6th October, the date deemed to be the expiry of the danger period for the *Carella*. The day after the Lords' meeting, a report and cover letter on the incident went to the Prime Minister, Churchill, assuring him that there had been 'no sequel' and 'the period of maximum risk is now over, although there is a slight and diminishing risk up until the 6th October. Precautions are therefore being maintained'.[41]

It was now October. The trawler and its naval stalker had reached Icelandic fishing waters without any sign of the crew falling ill. With the date and time of the expiry of the danger period firmly in place, the Admiralty and Ministry of Health maintained an anxious watch. *Truelove* was 'keeping a listening watch on the trawler's wireless waves, without shadowing or keeping in sight of the

---

36 TNA, ADM 1/26867. Signal (Top Secret). Admiralty to ZAMBESI (22 September 1952).

37 TNA, ADM 1/26867. Deputy Chief of the Naval Staff to First Sea Lord. Carella Incident (22 September 1952).

38 The Deputy Chief of the Naval Staff, formerly the Assistant Chief of the Naval Staff, was the most junior of the immediate assistants to the head of the navy, the Chief of the Naval Staff and First Sea Lord. For full details of the senior levels of the naval hierarchy at this time, see Grove (1987: 1–7).

39 TNA, ADM 1/26867. Deputy Chief of the Naval Staff to First Sea Lord. Carella Incident (22 September 1952).

40 Although not mentioned specifically in this correspondence, this meeting may have been of the Board of the Admiralty, where the Second Sea Lord was responsible for personnel and disciplinary matters. See Grove (1987).

41 TNA, PREM 11/251. Ian Jacob to Prime Minister (25 September 1952).

*Carella*'.[42] On shore, naval personnel made discreet enquiries of the trawler's owners, still maintaining the cover story that the crew might accidentally have seen secret material or weapons during the trial. These further enquiries at first revealed that the crew would return to Fleetwood around the 11th October; a day later the owners had revised this estimate to 6th October. In the interim, the Admiralty had already considered that if the trawler called in at a port prior to the end of the danger period, then some arrangements for keeping the crew under surveillance would need to be made.

Meanwhile, the *Truelove* tailed the *Carella* back from Iceland, keeping its distance at around 50 miles. And on October 3rd, with the *Carella* nearly home, a plan was laid out for the final few hours of the danger period. The waiting and watching policy remained in place, while the Prime Minister would only be informed of what was going on if 'anything untoward' happened.[43] Accordingly, a memo circulated to say that:

> The Ministry of Health has agreed to treat CARELLA on return to Fleetwood during the last few hours of the danger period on the same lines as they did when she was there before – i.e. no overt action, but specially briefed Ministry of Health official to go to Fleetwood with list of crew's home addresses, and see M.O.H. [Minister of Health] if necessary.[44]

Then, on 5th October, a brief and top secret signal was sent from the Admiralty to *Truelove*. It read 'cancel CARELLA operation. Ship has arrived in Fleetwood'.[45]

With the *Carella* in port and the official danger period over, the Admiralty and Ministry of Health were left to clear up the traces of their covert operation. On 6th October the letter of reprimand was sent to the commanding officer of *Ben Lomond*. At Fleetwood, in order to 'close and cover up [the] reason for our unusual interest', the *Carella*'s owners were informed that there had after all been nothing of interest on the deck of the *Ben Lomond* for the fishermen to accidentally spot.[46] Finally, a letter was sent out from the Admiralty to the Scottish Flag Officer on 16th October. The instructions in the letter placed a final seal on the whole matter:

> It has been decided that all the "evidence" about the recent incident with the BEN LOMOND and the CARELLA is to be destroyed with the exception of one complete record which will be kept in the Admiralty. To achieve this state of affairs, we are calling in all the papers and signals, and I therefore write to ask if

42 TNA, ADM 1/26867. Note by Admiralty to Chief Staff Officer to the Minister of Defence (2 October 1952).

43 TNA, ADM 1/26867. EML (initials). Trawler Carella (3 October 1952).

44 TNA, ADM 1/26867. EML (initials). Trawler Carella (3 October 1952).

45 TNA, ADM 1/26867. Signal (Top Secret). Admiralty to TRUELOVE (5 October 1952).

46 TNA, ADM 1/26867. EML (initials). Trawler Carella (3 October 1952).

> you would send to me personally BEN LOMOND's written report of which you had a copy… You will, I know, have already destroyed the three signals which we sent up to keep you posted; if you have any other papers which indicate what all the fuss was about, perhaps you would burn them too?[47]

## How Secrecy Operates, What Secrecy Permits

The immediate point to make about the *Carella* incident is that secrecy was woven into the fabric of the affair. Everywhere. Secrecy engendered further secrecy, as the initial secrecy of the operation was compromised and further actions to preserve secrecy had themselves to be performed in secret. In this way, covert behaviour and an accompanying silence spread, building on prior secrets and being pushed by the need to maintain secrets. Co-production occurred as the authority's attempts to maintain social order (secrecy) and produce knowledge (the monitoring experiment) operated simultaneously. Cauldron had been carried out under cover and the incident with the trawler meant that the monitoring exercise was put in place as a solution, in order to find out what had happened and to further maintain secrecy. Consequently, the experiment with the trawler crew depended on further concealment, from, for instance, the trawler owners, the crew of the *Truelove* and *Zambesi*, and the trawler crew if they had developed 'pneumonia'. Ending secrecy, by intervening and treating the crew with streptomycin, would also have ended the experiment in this particular form. Hence, secrecy became both an objective in itself and a means of carrying out the experiment.

Through the *Carella* incident we therefore see how secrecy operates in conditions where an experiment triggers an accident, and then blurs back into being an experiment. Here, we can see what secrecy allows in the productive, rather than restrictive, sense mentioned earlier. Secrecy certainly acted as a catalyst, shielding and permitting extraordinary transgressions in extraordinary circumstances. Levidow has previously noted that this blurring between accident and experiment also occurs in the wake of nuclear accidents (Levidow 1990). A loss of control during the accident is replaced with an on-going monitoring exercise that also generates data on the after-effects of the accident. This observation has certain similarities to the notion that, as there will always be uncertainties within science, society has become a laboratory for new science and technology (Beck 1992, Krohn and Weyer 1994, Strydom 2002, Szerszynski 2005).

With *Carella*, there was intervention (albeit unintended) and continued monitoring, including the generation of a secret protocol, which could have been stopped at any point by further intervention. So, while still straddling the boundary between accident and experiment, the events could quite justifiably be regarded as a form of deputed or vicarious experiment (Morgan 2003). Morgan uses this term to

47 TNA, ADM 1/26867. From Sandy Lombe to Rear-Admiral JHF Crombie (Flag Officer, Scotland). (16 October 1952). Ref. DCNS/428/52.

describe computer simulations and models; I am borrowing the term here to cover an investigation that stands in for a full-scale, planned experiment. Elsewhere, the term 'experiment of opportunity' has been used to capture this type of unplanned experiment (Moreno 2001). From a co-production perspective, the spanning of the accident-experiment boundary constitutes an attempt to restore social order, following the accident, by launching the secret monitoring experiment, which provided living confirmation of the limitations on the plague bomb's effects. Once again, problems of social and natural order demanded simultaneous solutions.

Additionally, under cover of secrecy several other ordinary categories blurred. The fishermen were simultaneously victims, citizens, experimental subjects and potential patients. Likewise, the navy and civil servants were also experimenters. There is no overt suggestion anywhere (as did happen with the Tuskegee syphilis trials) that failing to intervene with the trawler crew would provide interesting information. Ongoing monitoring, regardless of whether or not the crew became infected, would nonetheless have provided some information about exposure of humans to plague and, as mentioned, would have confirmed the limitations of the plague bomb. The scientists at Porton would undoubtedly have maintained a technical interest in the fate of the trawler crew, as evidenced by Morton's briefing on the situation. However, the navy and civil servants became the principal investigators as they took on the responsibility for trailing and monitoring the *Carella* and also for organising their contingency plan. If needed, they would also have been responsible for intervening to bring the experiment to a premature close. In more than a figurative sense, the patrons of the research became the researchers.

Throughout the incident, secrecy also acted as a way of orientating decision-making and behaviour. The course of action was dominated by the priority to maintain secrecy. This was obvious in the decision not to intervene by treating the crew with streptomycin and instead to maintain secrecy and security at all costs. It was also evident in subsequent decisions such as that of the Foreign Office not to send antibiotics to Iceland, not to fully inform the owners of the trawler, or the decision by the Admiralty not to hold a board of enquiry. Above all, it manifested itself in the Admiralty statement that 'grave political requirements are more important than what may seem best from a strictly medical point of view for individuals'.[48] In each instance, the metaphor of secrecy as simply the control over flows of information or as a tangled web ceases to be helpful. A more apt metaphor would be the labyrinth of secrecy, where detours and circuitous routes of behaviour have to be taken in order to navigate a way through. This also suggests, as was the case with the *Carella*, that maintaining secrecy made it difficult, if not impossible, for those in authority to reverse decisions about how to conduct their response to the incident.

In the culture of secrecy that pervaded Whitehall, it is not too surprising that the value of secrecy itself was not questioned in the written record of the incident.

---

48 TNA, ADM 1/26867. Signal (Top Secret). Admiralty to ZAMBESI (22 September 1952).

Although unquestioned, secrecy still had to be actively maintained. Beyond the obvious silences and limiting of information flow, the details of the case also suggest further mechanisms and social processes that sustained and encouraged secrecy. First, as events unfolded, secrets were treated to a greater extent as bounded entities. The bounded, contained secret had implications for the actions of the authorities as the secret became a peculiar form of synecdoche, where parts came to stand for wholes. As such, the contained secret implied that if part was released then everything else would be uncovered. This tactic was possibly indispensable for sustaining secrecy. It was certainly evident in the successive cutting back of information that would be contained in prepared press statements, with those in authority seeming to presume that even a small release of information would jeopardise the whole. It is equally apparent in the statement that killing rats onboard the *Carella* would point towards plague and effectively expose the whole of Operation Cauldron.

A corollary to the last point is that by treating the secret as a bounded entity, the decision-makers also had to attribute tremendous omniscience to those who might know the secret. We see such crediting of amazing powers in the example just cited of Sandys suggesting that if the authorities targeted the rats on the fishing vessel, then the crew would immediately know that they had been exposed to plague as part of a germ warfare test. Similarly, remarkable medical knowledge seems to have been assumed such that if plague had broken out on board the *Carella* the crew would recognise the seriousness of the early symptoms of fever in time to summon help. And the same case could be made for the imputed abilities of the press to uncover what was happening. Such 'pessimistic' estimates of how the public might gain knowledge stand in stark contrast to the 'optimistic' estimates that the *Carella* crew were probably unaffected by plague. All of which seems a variation on the usual conclusion from studies of the relationship between expert knowledge and the public, where those in authority are frequently shown to attribute to the public only ignorance and an inability to gain knowledge (Gregory and Miller 1998).

## Secrets and Science

It is a fairly straightforward matter to draw attention to the power differential established by secrecy. The military, politicians and civil servants remained in control of the exercise using a proliferation of secrets to stop the fishermen from knowing about their predicament. Secrets were also used to ensure the trawler crew remained oblivious that they were under surveillance. But, rather than simply saying that the whole affair was shrouded in secrecy, we catch a glimpse at how secrecy operates in practice.

Secrecy allowed a particular 'experiment of opportunity' to take place that spanned a boundary between accident and experiment. Secrecy and science were co-produced once this particular type of clandestine experiment was instigated as an alternative to the more open, but potentially more vulnerable, option of intervention. The option to monitor the trawler was made to appear significantly attractive, almost inevitable even, because the secret was treated as a bounded entity, with the implication that any small knowledge of the affair would breach security entirely. Then, in order to sustain the quasi-experiment, further secrets were needed to avoid others from discovering what had occurred. So, the isolated case of the *Carella* has implications for other forms of clandestine science. Once again, if we are to understand secret science, whether in a military context or elsewhere, we cannot conceptualise it merely as open science behind closed doors. A more nuanced understanding construes secrecy as more than that, more than a screen drawn around an otherwise untouched activity. Secrecy does not simply hide experiments; it is a social order which makes possible, and in turn proliferates through, different forms of knowledge production.

# Chapter 4

# Keeping, Disclosing and Breaching Secrets: Classification and Security

The *Carella* incident discussed at length in the previous chapter involved a certain degree of improvisation on the part of the authorities in order to maintain secrecy. Improvisation, in turn, took place on a platform of institutionalised secrecy. Coded signals, security classifications, cover stories and other routine measures all provided a repertoire of tools for 'doing' secrecy. A closer look at some of these mundane aspects of secrecy will help to think through two sides to the operation of secrecy: keeping secrets in and letting secrets out. Security classification, at its simplest a stamp marking out a document as, say, top secret or unclassified, is the focus in this chapter for understanding how secrets are contained. A close look at debates around classification of information reveals wider concerns at stake in the resolution of these discussions, notably how different national weapons research programmes might or might not coordinate and cooperate. Turning next to how secrets escape, I want to contrast the controlled release of information through press releases and other announcements with the uncontrolled breach of secrecy exemplified by leaks. Restoring control becomes a matter of the authorities employing various resources, including resources such as rumour and gossip, normally seen as the province of those who find themselves outside the sphere of secret knowledge.

## Keeping Secrets: Security Classification and the Value of Information

Scientists working within the British biological warfare programme were not entirely divorced from the world of civil science, including its reward system attached to publication and subsequent recognition. In 1953, researchers at Porton Down wanted to publish some of their work on immunisation against anthrax in the open scientific literature, a move opposed by their US and Canadian collaborators. Publication took on political significance, as scientific advisors in the UK recognised that publishing against the wishes of their collaborators would adversely affect relations with the USA. On the other hand, they pointed out, there was no obligation to suppress the work under existing security classifications. It was work deemed to be not sensitive enough to formally engender a Secret or Top Secret label. The advisors argued, moreover, that not publishing the work could imply that the UK was preparing for biological warfare; presumably this implication would only be made if the censure later became known about in public. The advisory board's 'technical opinion' at the outcome of this discussion was that the work should be published.[1]

The board's decision failed to resolve the situation, which dragged on for three months without resolution. During this time, the director of the UK biological warfare research programme, David Henderson, made personal approaches to the director of Camp Detrick, the key site of the US Army biological warfare programme, without any result. As an additional complication, two groups in the UK programme were now working on different approaches to producing anthrax antigen. Only one of the groups needed to refer to previous US work were they to publish, although Henderson foresaw problems with staff morale if one but not the other group was allowed to publish their findings. Eventually, through direct intervention by the US representative posted to the UK programme, the matter was resolved in favour of the UK.[2] This short-lived disagreement illustrates an argument that I want to advance in this chapter about security classification. A close examination of the historical record shows that classification and secrecy are not simply about the content of science. In this instance, scientific and political issues were distilled into a 'technical opinion'. Publications represented more than a means of disseminating scientific information, they were simultaneously an expression of the strength, or fragility, of collaboration between three nations.

Secrecy is enacted through a range of tools and practices, not least the classification of information. Classifying, of course, is a fundamental and ubiquitous activity through which we organise an otherwise disordered world. Yet the resultant taxonomies are far from neutral and inconsequential descriptions of the world, as sociologists Bowker and Starr point out in their detailed study of classification in action:

> These standards and classifications, however imbricated in our lives, are ordinarily invisible… Remarkably for such a central part of our lives, we stand for most part in formal ignorance of the social and moral order created by these invisible entities. Their impact is indisputable and… inescapable. Try the simple experiment of ignoring your gender classification and use whichever toilets are nearest; try to locate a library book shelved under the wrong Library of Congress catalogue number. (Bowker and Starr 1999: 2–3)

Classification therefore has a powerful role in creating and reinforcing order in the world. And whatever the reality of the situations they classify, taxonomies have very real effects. No less are security classifications a powerful way of ordering the world into varieties of secrets. Arguing along similar lines, White claims that secrecy is a way of valorising, or adding value to, information (White 2000). With security classification, this value is encoded in a hierarchy ranging from the most

---

1 TNA, WO188/668. AC12384/BRBM29. BRAB 29th Meeting (12 June 1953).

2 For a wider discussion of authorship and publication among defence scientists see Gusterson (2003).

valued, top secret, through to the least valued, unclassified, in relation to how particular powerful groups or individuals perceive the importance of any item of information. Turning back to the historical record there have been a number of protracted discussions about security classification and biological warfare research. Bearing in mind these observations that classification embeds values, creates order and has effects, we would be unwise to dismiss such disagreements over classification as mere bureaucracy. Instead, as we turn to the record of once secret government documents, we can ask: what else was at stake in the settlement of these disputes?

## Classification and UK–US Collaboration

Biological warfare security classification in the UK sat alongside a more general governmental system of security classification and vetting that had grown up during the Second World War. A four-tiered taxonomy, which formed the basis of the system later used throughout the Cold War, was adopted on 1 July 1943, namely: Most Secret, Secret, Confidential and Restricted.[3] The system was designed specifically to standardise UK and US systems of controlling sensitive information, and replaced the practice of labeling the most sensitive information 'to be kept under lock and key'. While this label had provided a clear directive as to how to treat the most sensitive secrets, its shortcoming was that it implied that documents not so marked could be left out when many departments already kept all their official documents locked away.[4]

The category of 'Most Secret' covered information about new methods of warfare together with their associated scientific and technical details, and handling of this information involved elaborate procedures. All documents at this level of sensitivity were to be transmitted 'by box (preferably black), in an envelope fastened with an economy label, wax sealed, and marked 'Most Secret – To be Opened by…'.[5] A list of people with permission to read it accompanied any most secret document. In this way, classification not only sorted information, but also people, into hierarchies of knowledge and knowing.

The term Top Secret soon replaced Most Secret. As the then Cabinet Secretary, Sir Edward Bridges, recalled nine years later, this ostensibly minor change facilitated a more fundamental alignment between the UK and USA, who had operated with just Secret, Confidential and Restricted:

---

3 TNA, CAB 21/1659. Security Classification of Official Documents (11 June 1943).

4 TNA, CAB 21/1659. RMJ Harris to Private Secretary. The 'Lock and Key Formula' (26 June 1943).

5 TNA, CAB 21/1659. Offices of the War Cabinet. Official Notice. Classifiation and Handling of Official Documents (26 June 1943).

> If my recollection is right, we had prolonged discussion with the Americans and found great difficulty in coming to terms with them. It appeared that their greatest objection was to the use of the term 'Most Secret', which – to their way of thinking – implied that papers which were called 'Secret' were not really secret at all.[6]

Bridges further explained that Most Secret 'referred so to speak to the papers which were kept in "the top drawer of the cupboard in which the secret files were kept".'[7] The replacement term Top Secret apparently appeased the USA by suggesting that secret files were indeed secret, and that Top Secret further denoted the 'top layer of secret papers'.[8] Although Bridges' anecdote again suggests a trivial change in nomenclature, there was a more serious underlying agenda: to coordinate work on both sides of the Atlantic.

The term Top Secret also indicated the serious consequences that could attend any unwarranted release of material in this category. By 1950, the US Joint Chiefs of Staff had explicitly incorporated these consequences into their definition of Top Secret. Loss of this information could result in one or more calamities: war against the USA, defeat in on-going or planned war operations, or 'loss by our nations of a scientific or technical advantage of sufficient military importance as to affect materially the course or outcome of a war or major operation'.[9] The serious nature of such loss was brought home at this time by fresh revelations in the UK that scientists had betrayed atomic secrets. Head of theoretical physics at the Harwell Atomic Energy Research Establishment, Klaus Fuchs, was arrested in January 1950 and shortly afterwards fellow physicist Bruno Pontecorvo escaped to the Soviet Union (Goodman 2004, Turchetti 2004). The British authorities were not only concerned about losing sensitive information, but were equally concerned about the impression now created that British security measures were lax (Goodman 2005). Hunting for spies had also become a commonplace security activity in the USA (Badash 2003). Against this backdrop, one historian of Whitehall secrecy characterised the atmosphere in the UK in the following terms:

> The conviction grew that those who consented to work in secret had secrets of their own to hide. The glad confidence in the value of science to the citizens of the welfare state was replaced by a less trusting and respectful attitude to those charged with discovering the hidden mysteries of nature.(Vincent 1998: 202)

Somewhat unexpectedly in this climate of distrust, a debate took place in Whitehall over the possible over-use of existing classificatory practices. Put succinctly by one

---

6 TNA, CAB 21/2836. Memo EEB to Sir Norman Brook (1 April 1952).

7 TNA, CAB 21/2836. Memo EEB to Sir Norman Brook (1 April 1952).

8 TNA, CAB 21/2836. Memo EEB to Sir Norman Brook (1 April 1952).

9 TNA, CAB 21/2836. From British Joint Services Mission (BJSM) Washington to Ministry of Defence (London) (27 April 1950).

civil servant: 'current security gradings are grossly abused. Top Secret has ceased to have any real significance'.[10] The problem of over-grading was accompanied by the sense that 'the distinction between security gradings… is in certain respects very fine. As a result individuals will always tend to allot a higher classification, preferring to err on the side of safety, where there is any element of doubt'.[11] Beyond conservatism, civil servants also complained that the proliferation of Top Secret material made communication difficult, made safe transit of documents difficult, increased the need for vetted staff, created storage problems, slowed down decision-making, added to administrative costs and that it 'increasingly defeats the object for which it is designed'.[12]

This antipathy culminated in April 1952, with Cabinet Secretary, Sir Norman Brook, writing to the Prime Minister, Winston Churchill, 'I share your dislike of the term 'Top Secret' and I wish that we never had to adopt it'.[13] Brook continued that it was, in fact, Churchill who had approved the term in 1944 in order to harmonise with the Americans, who had 'refused to employ the term "Most Secret" which we then used; our adoption of "Top Secret" was the price we had to pay for American acceptance of our four-tier system and standard definitions'. Presently, he added, the system was in such wide use throughout the Commonwealth and also other NATO countries, that it would not make any sense to change. Brook instead put several measures in place to ensure more restraint. A printed card with definitions of the security categories was made available to staff, a talk was given on 'security-mindedness', and staff were encouraged to challenge any incorrect classifications they encountered.[14]

## Anglo-American Coordination and the Classification of Biological Warfare Knowledge

Biological warfare security classification remained on the UK agenda as one particularly sensitive aspect of this broader concern with the practicalities of security. Again, classificatory battles were about the means of controlling information but also about coordinating UK research with the USA. During the early 1950s a protracted discussion took place between the two countries as they attempted to keep their research programmes co-coordinated even as the USA altered their classification schemes. Since the end of the Second World War, the

10 TNA, CAB 21/2836. Summary of Ministry of Defence Correspondence (1 August 1951).

11 TNA, CAB 21/2836. Cabinet.Inter-Departmental Committee on Security. Liaison Officers' Conference. Use of Security Classifications (2 November 1951).

12 TNA, CAB 21/2836. HL Verry (DSIR) to RJP Hewison (Cabinet Offices) (31 August 1951).

13 TNA, CAB 21/2836. Norman Brook to Prime Minister (3 April 1953).

14 TNA, CAB 21/2386. Use of Security Classifications (n.d).

UK had shared detailed classification schemes with the USA in just two areas, biological warfare and atomic matters.[15] In April 1951 the US Chiefs of Staff notified their UK counterparts that the BW Security Classifications that had held since 1948 were to be revised.[16] To illustrate the scheme, they provided a list of examples of what would now fall under each of the four categories from Top Secret downwards, so that the whole notification document ran to several pages in length.[17] The most contentious issue for the UK authorities was that some sensitive material would be downgraded to lower categories.

**Table 4.1 Proposed Changes to Security Classification Scheme for Biological Warfare agreed Between US and UK**

| Information | Proposed Downgrade From |
|---|---|
| 'The fact that specific living agents or their toxic derivatives, identified by name, code name and/or description have been selected for or are being manufactured for offensive military use' | TOP SECRET to SECRET |
| 'Full details of the processes involved in the manufacture of BW agents for offensive purposes' | TOP SECRET to SECRET. |
| 'Rosters and tables of organisation of BW installations and organisations' | CONFIDENTIAL to RESTRICTED |
| 'Information concerning methods of employment and tactical use of BW agents as necessary for defensive purposes' | RESTRICTED to UNCLASSIFIED |

*Note:* final downgrade to include the omission of the word 'tactical' and addition of the term 'for instruction and training in measures of defence against BW'.

*Source*: TNA, WO 188/665, BW(51)34. Chiefs of Staff Biological Warfare Sub-Committee. BW Security Classification (27 June 1951).

Within a month, the UK Air Ministry voiced objections, specifically to four of the proposed classification downgrades (see Table 4.1).[18] The Air Ministry argued that unwarranted access to the first two items, details of agents and manufacturing

15 TNA, WO 188/666, BW(54)5. Chiefs of Staff Committee Biological Warfare Sub-Committee. BW Security Classification (2 March 1954).

16 TNA, WO 188/665, BW(51)34. Chiefs of Staff Biological Warfare Sub-Committee. Security Classification of Biological Warfare Information. Report by Wing Commander GM Wyatt and Major DMC Prichard (26 November 1951).

17 TNA, WO 188/665, BW(51)34 Chiefs of Staff Biological Warfare Sub-Committee. BW Security Classification (27 June 1951).

18 TNA, WO 188/665. Annex. Copy of a letter received from DDI (ORG&S) dated 25/7/51 addressed to Wing Commander A.W. Howard, Ministry of Defence (25 July 1951).

processes, would help an enemy develop countermeasures. They then complained that revealing rosters and organisational tables would leave any named individuals vulnerable. The final item, an unclassified rating of material for training purposes, they thought would sacrifice control by moving this information outside the reach of the Official Secrets Act.

A further report for the high-level Chiefs of Staff's Biological Warfare Sub-Committee followed in November, in which the changes were all described as 'a considerable relaxation of Security Classifications'.[19] At the root of these disagreements were more fundamental differences between the secrecy cultures of the two nations. According to the report, and echoing the discussions described earlier during the Second World War, for the USA the clearest separation was between Secret and Confidential, whereas for the UK there was a particularly strong line drawn between Top Secret and Secret:

> USA tend to be much more sparing with the allotment of TOP SECRET classification than the UK... They [USA] tend to bracket TOP SECRET and SECRET together as a group well above the other Security Classifications, limiting TOP SECRET classifications to the minimum. The UK tendency, however, is to place TOP SECRET on a pinnacle apart and group SECRET and CONFIDENTIAL more together. As a result it is felt that SECRET has a higher significance in the USA than in UK.[20]

The differences, yet again, amounted to more than trivial differences in bureaucratic practice. As the last sentence in the quote suggests, the categories ended up carrying different significance and value within the two, supposedly harmonised, schemes. Classification reflected, or perhaps more accurately helped act out, different national policies and research priorities. The difference here is admittedly nuanced. Reflection suggests that there are first of all priorities and policies and then classification; acting out suggests that priorities and policies are nothing without their translation into mundane practical actions and objects such as classification and classification schemes. In this respect, the report noted that American pressure to relax security classification around BW came from the increasing number of people needing access to information as the US programme geared itself up for large-scale production of biological weapon agents: 'USA are far nearer to the potential production stage of BW agents than in the UK. Once

19 TNA, WO 188/665. BW(51)34. Chiefs of Staff Biological Warfare Sub-Committee. Security Classification of Biological Warfare Information. Report by Wing Commander GM Wyatt and Major DMC Prichard (26 November 1951).

20 TNA, WO 188/665. BW(51)34. Chiefs of Staff Biological Warfare Sub-Committee. Security Classification of Biological Warfare Information. Report by Wing Commander GM Wyatt and Major DMC Prichard (26 November 1951).

this stage is reached it is virtually impossible to maintain an unduly high Security Classification without imposing crippling handicaps on production planning'.[21]

Both countries operated a 'need to know' policy, but the number of people who now needed to know in the USA had grown. In the ensuing discussion among the UK Chiefs of Staff, this difference was repeatedly stressed and various officials pointed out that matters, such as the names of specific agents selected for BW, should remain top secret as long as possible, even if this entailed divergence from the USA.[22] The position of the UK was summarised by the Chairman of the Biological Warfare Sub-Committee, Lieutenant-General Sir Kenneth Crawford: 'We felt in general over the whole BW field that the amount of information given out should be the minimum and to as few people as possible, and that items should remain classified as highly as possible until it was absolutely necessary to downgrade them for production.'.[23]

Three years later, in 1954, there were further co-ordination problems when the USA decided to drop the term 'Restricted' entirely. Consequently, information classified in the UK as Restricted would, on arrival in the USA, be classified upwards to Confidential and therefore have a smaller distribution than intended by UK scientists and policy-makers. The Chief Scientist at the Ministry of Supply, Owen Wansborough-Jones, responded rather wearily that the UK should simply go its own way as 'little would be gained by having a long and tedious argument with the US. It would always be difficult to decide what classification to give any piece of information'.[24] The US Chemical Corps representative, present at the same meeting, differed on the grounds that 'since there was a great deal of information passed between the US and UK, it was very desirable that similar subjects were accorded the same security protection in the two countries'.[25]

The following year, when a draft of the 'US Chief Chemical Officers Guide to Security Classifications' was put before the UK Chiefs of Staff for comment, they reiterated their position on divergent classificatory schemes. While the differences were thought to have little practical effect, they nonetheless reflected deeper assumptions about the role and aims of the UK scientific programme compared with the USA. The Chiefs of Staff argument mirrored the argument about the production goals of the US programme. Security classification was different in the

---

21 TNA, WO 188/665. BW(51)34. Chiefs of Staff Biological Warfare Sub-Committee. Security Classification of Biological Warfare Information. Report by Wing Commander GM Wyatt and Major DMC Prichard (26 November 1951).

22 TNA, WO 188/665. BW(51)4th Meeting. Chiefs of Staff Committee, Biological Warfare Sub-Committee (2 November 1951).

23 TNA, WO 188/665. BW(51)4th Meeting. Chiefs of Staff Committee, Biological Warfare Sub-Committee (2 November 1951).

24 TNA, WO 188/666. BW(54)5. Chiefs of Staff Committee Biological Warfare Sub-Committee. BW Security Classification (2 March 1954).

25 TNA, WO 188/666. BW(54)5. Chiefs of Staff Committee Biological Warfare Sub-Committee. BW Security Classification (2 March 1954).

UK, according to the Chiefs of Staff and their scientific advisors, because the UK was pursuing more fundamental research:

> We appreciate that the position in the USA regarding research, development and production in the BW field is very different from that in this country where our effort is concentrated mainly on fundamental research. For this reason we realise that many of the examples given in the guide are governed by local conditions… The differences in the programmes and conditions in our two countries are such that there are bound to be differences in the detailed application of policy for security classification… For this reason we believe that any attempt to seek agreement on a guide as detailed as the one under comment would be neither desirable or possible.[26]

A short contribution by the Canadians to this debate also revealed a different problem created by a divergence in classificatory regimes. The fact that the USA was collaborating with Canada in their biological warfare research was proposed as an unclassified matter in the USA, yet revealing this collaboration would have had severe political repercussions for the Canadians who declared:

> Canadian policy does not permit the designation 'Unclassified' for Canadian work in the offensive field of BW agents primarily because of political considerations. This, of course, has no direct bearing on the statements in the document [draft Guide] as written but does mean that statements by the USA that they are participating in offensive BW work with Canada could not be considered 'Unclassified' by Canada.[27]

The Canadian concerns reinforce the point made several times so far, that much more was at stake in assigning and harmonising security classifications than the exercise of bureaucracy. For the USA, collaboration with Canada was unexceptional. On the other hand, Canada, with no offensive programme of its own but close collaboration with the US programme, could not afford for that information to be made public.

## Security Classification and Invisible Categories

A final point to make regarding the effects of classification schemes is that security classifications are powerful tools, not only for organising but equally for thinking. Some things are far easier to conceptualise within some classificatory

---

26 TNA, WO 188/666. COS(C)(55)61. From JH Gresswell, Joint Secretary, BW Sub-Committee to The Secretary, BJSM, Washington (26 May 1955).

27 TNA, WO 188/666. BW(55)9. Chiefs of Staff Committee Biological Warfare Sub-Committee. BW Security Classifications (27 June 1955).

schemes, conversely some things may be extremely difficult or even impossible to conceptualise using other schemes. In other words, beyond the silence of secrecy imposed by a classification scheme, the architecture of the scheme may create other ways of stifling information. This point is exemplified in the way that dual-use, the potential for the same science and technology to be used for peaceful or military purposes, was side-stepped in a 1952 US security classification system. In this system only matters deemed directly related to biological warfare were given a classification, whereas under 'unclassified' was listed:

> Technical and research data which are not of direct or anticipated military application , but are of interest and use to the arts, science, and public health.... For example, data presented as having been gathered in the normal course of non-military scientific research, unclassified military research, and data dealing with problems concerning natural forms of disease may be submitted for publication.[28]

All of this unclassified information could be published in the open scientific literature, with the names of the people and establishments that had produced the work.

This classification, which cleaves apart sensitive biological weapons research and work with no 'no direct or anticipated immediate military application', highlights a classificatory tension identified by Bowker and Starr (1999: 10–11). The categories of an idealised classification scheme, they argue, are mutually exclusive; in the real world, however, this ideal is hardly ever realised. Again, any classification scheme does not simply describe reality, it actively creates (and silences) categories for working with. The scheme quoted above for biological warfare security attempts to force the classifier to make an 'either-or' choice for any scientific and technical knowledge under consideration. It is, according to the quote, either a security problem or it is of benign interest to the 'arts, sciences and public health'. In this scheme, dual-use science would at once be Unclassified *and* Top Secret. It would, to use Bowker and Star's terminology, be a 'monster' transgressing the neat bureaucratic categories of the scheme. Or, put more prosaically, the classification scheme side-steps the problem of dual-use by classifying a world in which 'the dual-use problem' is simply assumed not to exist and, through the lens of the scheme, is rendered invisible.

## Letting Secrets Out: Orderly and Disorderly Release of Knowledge

If security classification imposes order on the world by organising it into a hierarchy of knowledge, the release of information to the wider public is also subject to

28 TNA, WO 188/666 BW(52)6. Chiefs of Staff Committee Biological Warfare Sub-Committee. BW Security Classifications. (21 April 1952) Annex. Policy Governing Classification of Matter Concerning Biological Warfare.

being carefully ordered and controlled. In previous chapters we have seen how press releases were carefully drafted and re-drafted, sometimes to be passed on to journalists and at other times to be held back as part of a contingency plan. A further example of this type of information management reveals how different assumptions are embedded into the text of a public announcement. Shortly after the end of the Second World War, the UK Chiefs of Staff considered a general public announcement about biological warfare, intended to be made with the US and Canadian authorities. In a report to the Chiefs of Staff Committee, advisors stated three reasons for making an announcement:

i. It would allay any anxiety on the part of the public through uncontrolled leakages;
ii. It would provide authority for the release of scientific information of value in industrial, agricultural and medical fields;
iii. It would assist in the integration of post-war biological research with Allied scientific studies.[29]

What was there to know? Primarily, that during the Second World War the UK developed a rudimentary anthrax weapon for use against livestock, and successfully tested an anti-personnel anthrax bomb (Balmer 2001, Carter 2000, Carter and Pearson 1999). Towards the end of the war, close collaboration with the USA and Canada resulted in the UK ordering, though never actually receiving, mass produced anthrax bombs from the USA. The public announcement on biological warfare read very differently. It explained that shortly after the fall of France 'it was decided that experiments should be carried out to assess with greater accuracy the possible effects of German secret weapons of this nature and to design defences'.[30] The announcement also reported that after 1942 collaboration with the USA and Canada was initiated in order to devote more resources to the work and 'to protect the country against any weapons containing bacteria which the enemy might produce'.

Here, the release of information is equally an effort to exercise control over secrets, with only carefully selected information made public. The announcement reveals assumptions made by its authors about the relationship between those who are party to secrecy and those outside. As the first point in the public announcement makes clear, it assumes a trusting public who would become undesirably anxious if they were contaminated with uncontrolled information leakages.

The second justification for a public announcement about biological warfare was to enable the free flow of information. The announcement, its authors argued,

---

29 TNA, CAB 80/51. Chiefs of Staff Committee. Public Announcement on Biological Warfare: Report by the Inter-Services Sub-Committee on Biological Warfare (7 November 1945).

30 In fact, it turned out that German attention to biological warfare had been little and uncoordinated (Geissler 1999).

would provide 'authority' for release of information to a wider audience. This phrase is not entirely clear but authority presumably would arise from bringing matters into the open. The announcement would create general public awareness, not only that biological warfare research had taken place, but also that it had serendipitously given rise to peaceful applications. In the words of the public announcement:

> These researches [into biological warfare] have resulted in the accumulation of much knowledge concerning fundamental problems in micro-biology, including new techniques applicable to the study of preventative medicines and of value to agriculture. It is the intention of His Majesty's Government to make the results of fundamental studies available to scientific workers generally, and arrangements for publication to learned societies and in appropriate scientific journals are now in progress.[31]

The 'authority' for further information release was also bolstered by presenting this case for the necessity of the wartime and post-war biological warfare research programme. Taking in the third benefit of the announcement, easing the Allied collaborative effort, the announcement stated that UK, Canadian and US research had 'revealed new techniques and new methods of approach which might constitute a potential danger to a country unprepared'. Consequently, the UK was to continue its biological warfare research programme because 'this is essential in order that means of defence against this form of war can be adequately planned and prepared, and action taken in the event of this form of warfare materialising'.[32]

Although the particular course of action to be taken remained unspecified in the document, shortly after the public announcement was drafted the UK laid plans to develop a strategic anti-personnel capability in biological warfare that would be equivalent to their nascent atomic bomb (Balmer 2001, 2006).

## Letting Secrets Out – Leaks

While security classification drew a firm line between classified and open knowledge, press releases and other public announcements allowed for controlled secrecy that contained carefully selected and worded information about the rationale and activities of the British programme. In contrast to all this orderly release of information and its attendant regulation of secrecy, past biological warfare programmes along with other secret weapons programmes, have faced

---

31 TNA, CAB 80/51. Chiefs of Staff Committee. Public Announcement on Biological Warfare: Report by the Inter-Services Sub-Committee on Biological Warfare (7 November 1945).

32 TNA, CAB 80/51. Chiefs of Staff Committee. Public Announcement on Biological Warfare: Report by the Inter-Services Sub-Committee on Biological Warfare (7 November 1945).

unplanned breaches of secrecy. This precarious aspect of secrecy was recognised by Shils, writing about the Cold War McCarthy witch hunts, he noted that 'one of the reasons why there is so much disturbance about the protection of secrets is that it is so extraordinarily difficult to do it well' (Shils 1956: 74).

This 'disturbance' can manifest itself as unforeseen effects, in Bratich's (2006: 507) words: 'It [secrecy] doesn't simply produce or create – it sends effects out into a future it cannot predict'. Such uncertainties can build an unusually tense and inward-looking atmosphere. In his careful analysis of spying allegations in the US nuclear research program at the end of the twentieth century, Masco observes that secrecy and the resultant fragmentation of information, where no one has the whole picture, contributes to just such an atmosphere of rumour and fear: 'Secrecy, however, is also wildly productive: it creates not only hierarchies of power and repression, but also unpredictable social effects, including new kinds of desire, fantasy, paranoia, and – above all – gossip' (Masco 2001: 451)

At Los Alamos in the late 1990s the accusations of Chinese spying and theft of information generated just such an atmosphere. Masco argues that this atmosphere was followed up with what he terms 'hypersecurity measures', with the introduction of unprecedented and baroque controls (for example, elaborate rules about who could associate with whom, or novel uses for technologies such as lie detectors) designed to either repair the breach in secrecy, or at the least allow managers to be seen to be managing the situation. Turning to biological warfare, Masco's observations can be extended by looking at two examples from the history of the British research programme.

The British biological warfare research programme was born in exemplary secrecy. Even Churchill, his War Cabinet, together with the most senior advisory committee on biological warfare, were not party to the initial decisions – engineered by senior civil servant, Maurice Hankey – to found a research group at Porton Down in Wiltshire (Balmer 2001). Once underway, the work was carried out behind the physical security of guards, gates and fences provided by the military research establishment. Records of the work, minutes of meetings, correspondence and memoranda were generally classified. Indeed, to the detriment of the historical record but the profit of security, much discussion and decision-making about biological warfare that took place in Whitehall and at Porton, at least until the end of the war, was carried out by word-of-mouth.[33]

By 1942 the physical protection of secrets offered by Porton Down was insufficient for the ambitions of the research programme. Progress towards a working anti-personnel weapon, the anthrax-charged 'N bomb', had reached the stage where scientists wanted to undertake outdoor trials beyond the bounds of Porton. The subsequent clandestine trials on the remote Scottish island of Gruinard are well documented (Carter 2000, Carter and Pearson 1999, Balmer 2001, Willis

33 More recently, word-of-mouth was also a way of maintaining secrecy and avoiding (later) accountability in the South African biological warfare programme (Gould and Hay 2006).

2004). Some months following the trials, security was threatened by an outbreak of anthrax affecting around 30–50 sheep and other assorted domestic animals on the nearby mainland. Paul Fildes, the head of the research programme, and his deputy David Henderson visited the site and attempted to stem the outbreak, apparently by setting fire to heather in the vicinity of the infections.

Fildes later noted to his superiors that the initial method for disposing of dead sheep from the trial, placing them with an explosive charge so that they would be buried under rocks, had gone awry. Too much charge was used and one or two sheep were blown into the sea, later to be washed onto the mainland shore, about three-quarters of a mile away. Rumours followed, but not entirely arising from public gossip. To begin with, the Security Services had been tasked with spreading a rumour that the outbreak had originated from a carcass that had dropped from a Greek ship at a nearby convoy assembly point. The rumour apparently took root, because it was soon followed up by a compensation claim to the Ministry of Agriculture. Fildes was informed of the claim and protested that allowing it would constitute a security breach. The Ministry of Agriculture representative concurred, proposing to decline the claim 'on the grounds that we are not responsible for animals thrown overboard from Greek ships'.[34] Eventually the compensation was granted along with an explanation that a settlement for the sum between Greece and the UK would most likely be made after the war.

We have already examined another security breach in chapter three, with the exposure of the crew of the fishing trawler *Carella* to a cloud of plague during Operation Cauldron in 1952. In the context of the present discussion, it is important to remind ourselves of some of the attempts made to prevent the exposure of the crew or wider public to the truth about their predicament. Rather than intervene with a precautionary dose of antibiotics, medics aboard the vessels sent to shadow the trawler were instructed, in the event of receiving a distress call, to board the ship and inform the crew that they were treating their stricken crew members for pneumonia. In the meantime, back on shore, Admiralty staff were instructed to approach the owners of the *Carella*. They were told to obtain the names and addresses of the trawler crew, plus details of any changes of personnel that might have occurred. The trawler's owners were told a cover story, 'that the trawler was in a forbidden area and that the men may have seen secret equipment which it was particularly desired that they should not'.[35] When it eventually transpired that the crew was unaffected by plague, the *Carella*'s owners were informed that there had after all been nothing of interest for the fishermen to accidentally spot.

In the context of this discussion of secrecy, these two incidents demonstrate how secrecy operates in a slightly different way to that proposed by Masco. In the Los Alamos case, Masco describes a security breach in a secretive atmosphere that generates rumour and gossip; this is then followed up by social control through the emplacement of hypersecurity measures. This leaves rumour and gossip in the

34 TNA, WO 188/654. Letter Fildes to Duff Cooper (14 April 1943).

35 TNA, ADM 1/26857. Letter Noble to R.A. Butler M.P. (18 September 1952).

province of those on the outside of the secret. A rhetorical division is forged between the rational, calming intervention of the authorities and the irrational, ill-informed set of ignorant outsiders prone to panic and gossip. Alternatively, the Gruinard and *Carella* incidents show how rumour and gossip actually meld with hypersecurity. In both cases, rumours were deliberately created by those in authority in order to control information and manage the situation. This is, of course, not uniquely a feature of biological warfare. Rumour and gossip are resources that were used by the authorities during the Manhattan Project. For example, there were deliberate attempts to spread a rumour in Santa Fe that an electric rocket was being built at Los Alamos (Serber 1998: 78–79).

At a broader level, this use of rumour and gossip alerts us to assumptions embedded in contemporary debates about security and the governance of biological weapons. In these discussions, rumour, gossip, not to mention paranoia, are often depicted as inherent attributes of 'others' (Williams 2004), for example an ignorant or irrational public. This is despite observations, made in the context of a discussion on the potential threat of bioterrorism, that the public can sometimes remain remarkably resilient in the face of a crisis (Durodié 2004). The examples from the historical record suggest that rumour and gossip are not simply something resorted to by those outside of the secret sphere, but can equally be construed as governance tools for keeping those 'others' ignorant.

This chapter has paused to reflect on the routine and mundane operation of secrecy in the British biological weapons programme. Security classification, in this case, reflected how British biological weapons scientists and advisors viewed the status of their own programme, both on its own terms – as having reached a particular stage of development and making fundamental research a priority - and in relation to its allies, in particular its standing with respect to the United States. Secondly, I pointed out that the release of secrets by the state is often accompanied by strong controls over just what is or is not revealed. While this observation is unexceptional, I have further argued that in cases where this control is lost we witness the authorities using tools not always immediately associated with democratic governance – including rumour and gossip – to restore order and maintain a grip on secrecy. Throughout, I have emphasised that, although the publication or otherwise of scientific information is important, more than just this fate can be at stake when keeping and revealing secrets.

# Chapter 5
# Secrecy, Doubt and Uncertainty: Power/Ignorance?

The truism that scientists pursue the creation of new knowledge can distract us from a significant characteristic of the relationship between secrecy and knowledge: the role of doubt, uncertainty and ignorance. Secrecy disrupts knowledge, some people possess secrets and others are left outside, not knowing. Even this formulation is too neat, as we shall see in this chapter which turns from our focus so far, the relationship between secrecy and knowledge, to secrecy and uncertainty about science in the policy-making arena. Experts inside the sphere of secrecy also make use of ignorance and uncertainty. Conversely, those outside of a secret will often make use of guesswork and inference in order to try and fill in their knowledge gaps (Rappert 2009). My main aim in this chapter is to demonstrate how scientists acting as expert advisors draw on their authority not only to lay claim to certainty, but also to uncertainty. Put bluntly, scientists are socially legitimated doubters. While this claim is fairly straightforward, I will argue that it has largely been overlooked in the academic literature that focuses on understanding the nature and role of science. Within the field of Science and Technology Studies, uncertainty is regarded almost entirely in negative terms: as something which scientists in policy disputes avoid at all costs, or utilise solely in order to deconstruct the claims of their opponents.

The chapter opens with an extended consideration of ideas about uncertainty and ignorance in science; it draws on recent work in Science and Technology Studies which takes us a little way from chemical and biological warfare. Later in the chapter the focus returns to the expert advisors in the history of the British biological warfare programme after the Second World War. We have already seen that many of Britain's secret policies since 1945 in this area were formulated by the joint Chiefs of Staff with direct input from scientific advisors. Indeed, this was a period where civilian scientists were routinely employed in the provision of scientific advice for the military (Edgerton 2006). As Britain moved from an offensive to a defensive stance over biological warfare, the status of biological warfare as an opportunity and threat, the status of research on biological warfare, and the significance of scientific advisors in the policy process all varied. Much of the time, the expert advisors on various committees acted as would be expected from a reading of the social studies of science literature. Expert advisors presented information on biological warfare as factual information which ostensibly spoke for itself with straightforward, direct and unequivocal consequences for policy.

The status of the advisors and their scientific constituency, by which I mean the areas where their input was deemed legitimate, were thus reinforced.

The same advisors, however, were quite capable of weakening their own claims without assistance. Experts frequently portrayed the findings of biological warfare research to their audiences as uncertain, in dispute and with negotiable consequences for policy. When successfully deployed, the effect, rather than diminishing the status and autonomy of the experts and their constituent scientists, was to present the field as one demanding more research, more resources and a higher status in policy deliberations. These activities, where scientists confess their ignorance and uncertainty, appear to have been left out of most sociological studies which present 'certain science' as needing a helping deconstructive hand.[1]

## Uncertainty and Expertise in Science Policy

Uncertainty and indeterminacy are no strangers to sociologists of science. Laboratory studies have thrived on the juxtaposition of certain knowledge produced by scientists and the processes by which pervasive uncertainties are removed in the course of knowledge production (e.g. Knorr-Cetina 1981, Latour and Woolgar 1986, Latour 1987, Star 1985). Finished science is represented, in these accounts, as having a radically different demeanour to unfinished science. In Latour's terms: 'the science made, the science bought, the science known, bears little resemblance to science in the making, science uncertain and unknown' (Latour1991: 7).

In the policy arena, scientific certainty carries particular authoritative weight when agencies attempt to justify particular decisions or courses of action. The stamp of science is frequently contested as opponents of any 'scientific' decision can equally depict it as politically motivated (see, for example, Collingridge and Reeve 1986, Englehardt and Caplan 1987, Jasanoff 1987, Nelkin 1992). One key process by which any activity or claim becomes designated as scientific or otherwise has been termed boundary work (Gieryn 1983, 1995, 1999; Jasanoff 1987, 1990). While boundary-work is frequently used within the scientific community to draw distinctions between science and pseudo-science, it is equally used in the policy arena to separate credible from non-credible experts, and scientific from policy concerns.

When scientists providing expert advice are successful at boundary work, their uncertainties are banished to the opposite side of the boundary under the rubric of 'policy' or 'policy science'. On the other hand, when policy-makers or opposing groups of scientists succeed, the soft underbelly of uncertainty is exposed to make

1 This theme can be compared with Lynch's (1998) observation that lawyers in the O.J. Simpson trial did an equally good job of deconstructing scientific claims as any competent ethnographer of science. With the scientific advisors discussed in this paper the expense of lawyers or sociologists or opposing scientists is spared when scientists flag the contingencies in their own claims.

data appear less than authoritative.[2] Jasanoff (1987: 224) summarises what is at stake in any such situation: 'For scientists, the primary interest in these boundary disputes is to draw the lines between science and policy in ways that best preserve the authority and integrity of science'.

What counts as scientific is therefore socially negotiated through boundary work activities. To take a fairly recent example from science policy, during the BSE or 'mad cow disease' crisis of the 1990s in the UK the question 'is beef safe to eat' was a central issue. In abstraction, this could be construed as either a scientific matter – could this new disease of cattle be passed to humans through eating beef? – or a policy matter – what risk is the government prepared to take? But in practice, and at different times by different groups, the two issues were variously conflated or separated, and were claimed to be the responsibility of scientists, policy-makers, or both (Millstone and van Zwanenberg 2001, van Zwanenberg and Millstone 2005). Rather than seeing the world fall neatly and without aid into the boxes 'science' and 'policy', it is this dynamic process of portioning out of what is regarded as the province of science, and what is regarded as the province of policy that can be seen as boundary-work. What is at stake in the resolution of any boundary dispute is the group's more general claim to legitimately access material and symbolic resources such as funding, authority, autonomy, trust and their ilk. In the context of this chapter, the main point of interest is that boundary work is a key mechanism for generating a representation of science as deliverer of certainty into the policy-making process.

A corollary of such disputes is that certainty and uncertainty become socially distributed. In this respect, although not in relation to boundary work, MacKenzie (1990) presents the 'certainty trough' as an extremely useful heuristic for understanding this distribution. From his study of the sociology of ballistic missile guidance, he argues that scientists and engineers close to the site of knowledge production and those distant and opposed to a technology are groups who will admit or draw attention to uncertainties and doubts. Between these extremes, users who are committed to a technology will fall into the 'certainty trough' and gloss over uncertainties that are apparent elsewhere. As MacKenzie (1990: 371) phrases it, they 'believe what the brochures tell them'. From here, it is only a short conceptual step from painting the distribution of uncertainty as relatively fixed and to interpreting it in more dynamic terms as constructed. In other words, where someone falls in the distribution of uncertainty may not always be so readily determined from their location relative to the site of knowledge production. Building on this idea of distributed uncertainty, not everyone has the same licence to doubt and that warrant will be valid only in certain places. So, to illustrate this point, from the view *within* the uncertainty trough, the uncertainties expressed by weapons designers are likely to be taken more seriously than those expressed by

---

2 Unintentionally, and somewhat paradoxically, this implies a thoroughgoing constructivism about certainty but a naive realist perspective on uncertainties which are 'there' in the data waiting to be identified.

protest groups. Within the protest groups, on the other hand, they are likely to take the uncertainties that they identify very seriously indeed.

Despite this pervasiveness of uncertainty as a theme within sociology of science, there is little analytic focus on ignorance itself. Knowledge remains the prime focus and uncertainties are usually an adjunct to studies of the construction and deconstruction of facts. Terms such as ignorance and uncertainty are used often used interchangeably in the few studies that do focus on their nature and role in science. The general term, incertitude, has been adopted by Stirling in a useful schema which aims to demonstrate the limits of a probabilistic risk paradigm (Stirling 2007). Whereas classic risk analysis regards risk as calculable from the magnitude of an event multiplied by its frequency of occurrence, Stirling points out that this situation, where both these parameters are known, is not a universal occurrence. Particularly where novel science and technology is involved, with possible unforeseen effects in the short or long term, then no one is in a position to specify either the probability of an event or the magnitude and nature of effects or both. In these situations, Stirling argues that we are in situations of, respectively, uncertainty, ambiguity and ignorance. This is a move foreshadowed by Keynes (1937), who distinguished probable knowledge, for example the chances of winning a lottery, from uncertain knowledge, such as interest rates in twenty years' time, where there is no scientific basis on which to form any calculable probability whatsoever'.[3] For Ravetz and Funtowicz (1993), when we are dealing with these types of incertitude, that escape probabilistic thinking or even the more tacit skills of the expert consultant, we are dealing with what they term 'post-normal science'.

Running through these various attempts to classify incertitudes is the attempt to delineate a completely ineffable notion of ignorance from false knowledge or from topics where actors can attempt to specify what needs to be known but remains unknown. Using terminology slightly at odds with the previous discussion, sociologist Matthias Gross has recently produced a useful analysis of this aspect of ignorance and uncertainty (Gross 2007). Gross reserves the term nescience, the want of knowledge, for what has been termed ignorance in the previous discussion; nescience is the completely unknown and unknowable. Ignorance, on the other hand is specifiable, and involves 'attempts to circumscribe the unknown' (Gross 2007: 743) to provide 'knowledge about the limits of knowledge' (Gross 2007: 751). The utility of this formulation is that it draws attention to the activity involved in constructing ignorance, someone must be doing the circumscribing and specifying, and here is a possible role for doubt in science and in the policy-making process.

Doubting, circumscribing, specifying are all activities amenable to sociological analysis, and this resonates with Smithson's (1989) observation that uncertainty and ignorance should themselves be treated as socially constructed. While undoubtedly useful, the Stirling definitions and Gross' definitions do not

3 For a full discussion see Gilles (2003). I am grateful to Donald Gilles for drawing my attention to this link between economics and current thinking on risk.

entirely match up with each other, and because it makes little additional difference to my main argument about who gets to express and construct ignorance, I will continue to use the terms interchangeably. The more general point to emerge from considering their perspectives is that 'ignorance, like knowledge, is not (simply) a 'given' in people or nature, but is (at least in part) a construction embedded in diverse social interests and commitments' (Stocking 1998: 168)

Likewise, Wynne has pointed out that in a world of pervasive uncertainties, 'uncertainty becomes a problem when we *interpret* it as a problem' (Wynne 1987: 95). Conversely, I would add, uncertainty only becomes an advantage when actively interpreted as such. Adopting this formulation shifts analytical attention to how uncertainty is assembled and used in practice. Uncertainty, from this perspective, can usefully be construed as a discursive resource for building up or destroying expert credibility (Michael 1994).

As within sociology of knowledge-focussed investigations, uncertainty is most readily construed by sociologists of ignorance in the 'destroying expert credibility' sense. It is something negative and therefore to be deployed as an offensive resource. A good example is Stocking and Holstein's study of the role of journalists in helping a trade association to 'manufacture doubt' about the findings of a critical study of the pig farming industry (Stocking and Holstein 2009). Along similar lines, Proctor discusses the deliberate manufacture of doubt by the tobacco industry concerning the hazards of smoking (Proctor 2006). In such cases uncertainty is something not to disclose because it is assumed that admissions of uncertainty will invariably weaken an actor's credibility. As Ravetz (1993) has put it: 'In the classic image of science, purveyed by philosophers and publicists, and imbibed by generations of teachers and their pupils, science is about certainty. Uncertainty is to be banished, and ignorance is to be rolled back beyond the horizon'.

And, insofar as sociologists have taken uncertainty to be something hidden in public accounts of science, they could be added to the list of philosophers and publicists. If uncertainty is to be banished then scientists need to avoid or manage it, for example through their use of diversionary caveats or emphasis on prior successes. Evans (1997) has demonstrated how economic forecasters employ methods for managing the large uncertainties in their models, for instance they use their past performance as an indicator of the robustness of their models. This permits them to admit the uncertainties without fully disclosing them. Rier (1999), to take another example, argues that the caveat sections of epidemiology papers can be used for maintaining their authors' credibility by pre-empting problems that might be raised by other scientists and the public. If uncertainty is regarded in this manner, as pollution to escape from, then it becomes a label which scientists and policy-makers may attempt to pin to their opponents rather than on themselves. Smithson suggests that, in this respect, 'ignorance becomes strategically important when a case needs to be made for stopping an activity... or against changing the status quo' (Smithson 1989: 240). A good example is Campbell's study of expert disputes in an oil pipeline inquiry where he emphasises how uncertainties were attributed to science by those scientists opposed to the pipeline and ignored or understated by pipeline supporters (Campbell 1985). In this case, uncertainty as a symbolic resource was minimised by actors concerned with their own credibility and emphasised by scientists wishing to undermine others' credibility.

Yet, ignorance and uncertainty may not always be such unwanted guests. They may be used in Michael's second sense described above, the 'building up expert credibility' sense. Hence, Wynne has pointed out without expanding far on the point, that 'sometimes institutions and people embrace uncertainty... in furtherance of their legitimate roles' (Wynne 1987: 95). When increased funding is at stake, for instance, scientists may be particularly keen to draw the attention of their potential patrons to knowledge gaps and how further research is needed (Stocking and Holstein 1993). Shackley and Wynne (1996) have drawn attention to this alternative deployment of uncertainty. They observed how scientists, in policy discussions over global climate change modelling, did not hide their own uncertainties and instead freely drew attention to them. However, in what they term 'boundary ordering', this indeterminacy was generally represented as tractable and manageable by these self-confessed uncertain experts. As a consequence, the expert advisors retained credibility with both their peers (who saw that they were not oversimplifying matters) and their patrons (who were presented with manageable uncertainty).[4] A similarly positive self-assignation of uncertainty is noted by Zehr (2000) in his study of press coverage of global warming and climate change. He notes that scientists openly refer to uncertainty in press interviews but adds that journalists use this uncertainty to draw a distinction between the considered caution of science and the 'hysterical' public wanting to draw conclusions too hastily. Zehr argues that journalists thus used uncertainty to perform boundary work and construct stereotypical identities for both experts and laypeople.

These ideas have been pushed further by McGoey in her analysis of pharmaceutical regulation in the UK (McGoey 2007, 2009). McGoey demonstrates how regulators at the Medicines and Healthcare Products Regulatory Agency (MHRA) deployed ignorance as what she terms an 'anti-strategy', where the ostensible aims of the regulators to provide knowledge about adverse drug reactions are constantly subverted by the pull to not provide definitive answers (McGoey 2007). This pull, she argues, arises from the need to maintain good working relationships with the pharmaceutical industry, who were also the agency's funders, and also to their claim that revealing too much information might scare the public and damage public trust in the whole regulatory system. This active sense in which uncertainty is put to use to produce effects leads McGoey to suggest that uncertainty is both generative and performative (McGoey 2009).

In all these studies of environmental and regulatory scientists, what might be dubbed 'confessional uncertainty' becomes a positive, defensive resource. Such expressions of uncertainty are not just features of contemporary science operating

---

4 Wynne (1995) has also argued that lay people can actively construct their ignorance of science and justify why they do not want to acquire technical knowledge.

in a climate of public ambivalence and scepticism, but can be found during the relatively more deferential period immediately after the Second World War – in the historical record of the British biological warfare programme.[5]

## Secrecy, Uncertainty and Ignorance

Secrecy is active in the construction of ignorance. Connections between secrecy and ignorance were explored by Simmel (1906) in his seminal paper on the secret, referred to earlier in this book.[6] Social interaction, according to Simmel, depends on knowing something about another person. Equally, interaction inevitably takes place when quite a lot else is not known about a person, some of which needs to be inferred in order to successfully conduct social life. For Simmel, concealment and, in its bluntest form, lying, were fundamental ways in which this balance between knowledge and ignorance could be deliberately manipulated. Simmel moves from this analysis of face-to-face interaction to examine the social dynamics of collective efforts by secret societies to conceal knowledge. His focus at this point is on insider-outsider relations, whereas I want to draw attention to relationships of ignorance and knowledge within secret communities. The most obvious way that the landscape of these communities is divided into knowledge, uncertainty and ignorance is through compartmentalisation as discussed in the opening chapter. Even without compartmentalisation, elite access to technical knowledge forges a divide between expert scientific advisors and their policy-making audience.

I will now examine some of the most telling confessions of uncertainty and ignorance in the open records of the Cold War British biological warfare programme. Just like the cases I have discussed involving uncertainty claims, those of the expert advisors were deployed as strategic, discursive resources. In this instance, the immediate goal was to perpetuate an offensive biological warfare research programme. As such, experts were able to 'cash in' the ostensible credibility of science so as to secure support. Yet, as will become apparent, uncertainty claims were not advanced in isolation but rested on wider claims to certainty and neutrality. Moreover, uncertainty – like certainty – is contestable ground and proponents of the research programme had variable success in employing it to persuade the authorities of the significance of biological warfare.

---

5 The notion that contemporary science operates in a more sceptical, ambivalent social context than earlier in the twentieth century is quite pervasive. At a broadly theoretical level it is associated with late modern and postmodern societies. See for examples Giddens (1990), Beck (1992), Crook, Pakulski, and Waters (1992), Lyotard (1984), Toulmin, (1991). An empirical demonstration of public ambivalence toward science is Evans and Durant (1995).

6 The link made by Simmel is also revisited in Gross (2007) and Marx and Muschert (2008).

## The Advisory Network

Within a year of the end of the Second World War a number of advisory committees had been established to advise on the direction of the UK biological warfare programme and – of equal importance – to channel research findings into deliberations about policy. Three committees exerted a substantial influence over the development of biological warfare research and policy. The Defence Research Policy Committee (DRPC) was located in the Ministry of Defence and had responsibility for setting priorities across the board for most defence research and development (R&D) programmes (Agar and Balmer 1998).The Chiefs of Staff Sub-Committee on Biological Warfare drew its membership from senior military, civil service and scientific experts. Their remit was to discuss and formulate all aspects of biological warfare policy. Located within the Ministry of Defence, the sub-committee reported to the Chiefs of Staff Committee who, in turn, reported to the Cabinet Defence Committee. Within the Ministry of Supply, the Biological Research Advisory Board (BRAB) consisted mainly of independent scientific advisors. These were senior figures drawn from various universities and research institutes from around the country. The BRAB's role was primarily to provide advice on the running of the MRD but was frequently consulted on technical matters by other committees.

A range of committees thus formed a chain of advice and guidance from the laboratory through to the Government, in principle with a division of labour between policy and science. Under the guidance of these various committees, the next decade saw the priority of biological warfare peak and then sharply decline as Britain moved from an offensive policy, aimed at creating a militarily useful anti-personnel biological bomb, to an entirely defensive policy regime (Balmer 2006, 2001, 1997).

### *Fear of the Possible*

The early days of the advisory committees witnessed considerable debate over the status of biological warfare. What could be achieved over the next decade? Should policy be offensive or defensive? Where did biological weapons lie in relation to other military priorities? Expert assessments of germ warfare soon began to trade on the destructive potential of biological weapons. But equally, and somewhat in contradiction, they also flagged uncertainties that any weapon could ever fulfil that dreadful potential. This tension generated a rationale that helped place biological weapons research at the top of the agenda: remaining uncertainties would yield to further scientific investigation. The scientists on the DRPC argued this point to the Chiefs of Staff in 1947, noting that:

> Biological weapons are potentially of greater value than chemical weapons and possibly of equal value to atomic weapons for breaking the will of any enemy to fight. *But plans should not yet be influenced by the prospect of the availability*

> *of such weapons within ten years. There is far too little knowledge available yet to justify this.* The whole subject needs the most intensive study, on a far larger scale than is being attempted at present.[7]

Here a claim about the potential of biological warfare is followed first by a statement of uncertainty qualifying the claim, and then a promise to resolve the uncertainty through further research. This tension between what was still unknown and what was feared possible was not confined to the DRPC and pervades much of the deliberation surrounding biological warfare at this time. The conclusion of the 1947 Annual Report by the BW Sub-Committee to the Chiefs of Staff echoed the same uncertainties through a contrast with atomic warfare and is worth quoting at length:

> BW [biological warfare] is a method of warfare arising out of the normal development of science and is in this respect equivalent to the atomic bomb. It differs from the atomic bomb in one major respect. It is untried. Given that the atomic bomb will function, it follows inevitably without experiment that a blast will be produced destructive to life and property. A BW weapon may demonstrably function, but it is impossible to estimate what effect it will have on life (man) because experiments cannot be carried out on man. Thus, all anticipations and deductions made in the present report are to be accepted only as sound conclusions based on experiments with animals, it being recognised that results on animals cannot with certainty be applied to man because the necessary quantitative data are lacking and cannot be gathered. Nevertheless it would be hazardous to assume that the estimated danger to man is less than has been concluded in this report. Expert opinion would certainly not support the assumption.[8]

Here, within a single paragraph, was a 'scientific' statement of the magnitude of the threat – equivalent to the atomic bomb – set alongside a commentary on the uncertainty of the threat. Doubts would always pervade any scientific extrapolation of results from non-human animal experiments to humans. The juxtaposition strongly implied that more research was imperative if the two positions (i.e. the magnitude of the threat, but clouded by uncertainty) were ever to be reconciled, although this could never be a complete reconciliation until a weapon was tried in combat. Significantly, this justification of research in terms of uncertainty and the 'fear of the possible' was set to become an influential motif.

Having also drawn on this 'fear of possible' line of argument, the DRPC recommended in May 1947 that 'research on chemical and biological weapons

---

7 TNA, DEFE 10/18. DRP (47) 53.DRPC. Future Defence Policy. 1 May 1947. Emphasis added.

8 TNA, WO188/664. AC10911/BRB83.BRAB. Report on BW Policy (3 May 1950). Appendix. Extract from 1947 report on Biological Warfare BW (47) 20 Final.

should be given priority effectively equal to that given to the study of atomic energy'.[9] In a later policy statement they added that a biological weapon 'may eventually prove comparable with the atomic bomb'.[10] Elsewhere, the BW Sub-Committee, in a document primarily expressing concern over the shortage of researchers in biological warfare research, had already reported that: 'We have been advised by the Deputy Chiefs of Staff Committee that research on BW is of the highest priority and in the same category of research as atomic energy and research'.[11]

With biological warfare elevated to the same status as atomic weapons, the practical outcome was a DRPC recommendation that the level of research on biological warfare should be increased. As described in earlier chapters, some £2.5m was spent on building extensive new facilities at Porton Down, Wiltshire and at the same time an ambitious programme of laboratory studies and open-air field trials was initiated. In addition to the agents investigated during the war, anthrax and botulinum toxin, the programme broadened in scope to consider pathogens responsible for diseases such as brucellosis, tularemia and plague. The next few years saw a general expansion of the entire programme in terms of personnel and resources as the topic enjoyed a brief period alongside atomic warfare at the pinnacle of British defence research policy.

## Mass Destruction

After the Second World War, chemical and biological weapons, together with atomic weapons, were classified by the United Nations under the rubric of weapons of mass destruction (SIPRI 1971,Vol. 4). By no means the first use of the term, for instance it had been used to refer to the strategic bombing of Guernica during the Spanish Civil War, a UN General Assembly resolution of 24 January 1946 called for the elimination of atomic weapons and all other weapons of mass destruction. Debate continued within the UN as to what else, if anything else, was to be included in this category. The UK policy, which equated biological with atomic weapons in terms of their priority, was not universally interpreted as meaning that both were in the same category of destructiveness. At various times between 1946 and 1949, statements can be found by the Chiefs of Staff and their advisors contesting that biological weapons both were and were not to be classified alongside atomic weapons as a new class of 'Weapons of Mass Destruction'.

The Chiefs of Staff's own planning staff had stated as early as 1946 that: 'it may be argued that almost any form of weapon, if employed in sufficient quantity

9 TNA, DEFE 10/18. DRP (47) 53. DRPC.Future Defence Policy.(1 May 1947).

10 TNA, DEFE 10/19. DRPC. Final Version of Paper on Future of Defence Research Policy (30 July 1947).

11 TNA, WO 188/660. ISSBW. Shortage of Scientific Staff for Research in Biological Warfare (30 January 1947).

and against suitable targets, is capable of mass destruction'.[12] They added that this clearly rendered the category far too broad to be useful, particularly in discussions about disarmament, but if used should be restricted to atomic and biological weapons. A year later, the Chiefs of Staff agreed on 'a cardinal principle of policy to be prepared to use weapons of mass destruction'.[13] Here, weapons of mass destruction were taken to be atomic, biological and chemical weapons.

By May 1948, however, the BW Sub-Committee reported to the Chiefs of Staff that: 'we do not suggest that Biological Warfare should be rated now or in the near future as a means for mass destruction of life'.[14] And a month later, the Chiefs of Staff were presented with a detailed US report which also suggested that biological warfare could not be considered a weapon of mass destruction. Likewise, the British Air Ministry continued to suggest that biological bombs would not be weapons of mass destruction. A cover letter for the Air Ministry's contribution to the 1949 Biological Warfare report stated that 'it is not suggested that Biological Warfare can be rated now or in the near future as a weapon of mass destruction but undoubtedly progress is likely to increase rather than reduce this fear'.[15]

This assessment of the potential of germ warfare acknowledged that the status of biological weapons was open to revision and indeed, after this time, few other doubts appear to have been raised over the status of biological weapons as weapons of mass destruction. The transition seems to have been made not through any explicit decision to bring the two into the same taxonomic fold, but rather through the UN deliberations and, within the UK, an increasing number of reports which, while heavily qualified, attempted to compare the potential effects of the two weapons. All this is significant because once the classification began to settle, then the fate of biological weapons became closely tied in with that of atomic weapons. As the policy community became increasingly enamoured with nuclear weapons in the 1950s, their priority was elevated above that of biological warfare.

## From Offence to Defence

The initial signs that faith in the powers of biological weapons might be faltering appeared in discussions over a 1949 report by BRAB to the Chiefs of Staff. The Board members continued to flag uncertainties but attempted a comparison with

---

12 TNA, DEFE 2/1252. Chiefs of Staff Joint Planning Staff. Control of Atomic Energy. Report by the Joint Planning Staff (27 February 1946).

13 TNA, DEFE 10/19. DRPC.Final Version of Paper on Future of Defence Research Policy.(30 July 1947).

14 TNA, WO188/662. Annex to BW(48)9(0)Final or COS(48)108(0). BW Sub-Committee Report to the Chiefs of Staff on the attitude to be adopted by representatives of the Atomic Energy Commission. (10 May 1948).

15 TNA, AIR [Ministry] 20/11355. 1949 Report on BW. BW(49)40. 2ndDraft (27 January 1950). Note from W.F. Lamb.

chemical rather than atomic weapons to draw attention to the destructiveness of biological warfare:

> An accurate assessment of the potentialities of Biological Warfare cannot, of course, be made until this weapon has been used on a large scale in warfare, and it may well be that when the actual trial is made against man – if ever it should be – it may prove relatively ineffective if not completely so. Nevertheless, a study of the potentialities of BW agents by the only available means (i.e. by trials on experimental animals) has shown that agents can be dispersed from aircraft bombs, and that the toxicity of three agents at least is many times greater than that of any known chemical agent.[16]

Here, the uncertainty claim that biowarfare might eventually prove ineffective, was balanced by a firm statement of the superiority of biological over chemical weapons. Unlike previous scientific declarations of ignorance, which were linked to calls for continued offensive research, the DRPC now interpreted this uncertainty surrounding the possible success of the weapons programme negatively. At a BW Sub-Committee meeting in May 1950 the dismissive views of the DRPC concerning the 1949 Report on Biological Warfare were reported that 'doubts had been expressed whether there was sufficiently convincing evidence on which to base a decision to make a special effort to continue research and development in the offensive sphere, or that Biological Warfare would be used against us'.[17] The new chairman of the BW Sub-Committee, Lieutenant-General Kenneth Crawford, tried to rescue the report from any such negative interpretations. He pointed out that estimates about the practicability of a biological weapon were 'the honest opinions of our technical advisers. Scientists by nature tend to be conservative, and the Report, may give a misleading impression that there are very long odds against BW ever being a practical proposition'.[18]

In the light of this disagreement, the DRPC asked Crawford for an update on the advantages and disadvantages of offensive biological warfare research. Crawford, in turn, consulted BRAB 'as a body with full scientific background and as men of common sense' and asked them to justify biological warfare, not so much in terms of economics, but against the likely use of atomic bombs as the major strategic weapons of the future.[19]

---

16 TNA, WO188/664. AC10911/BRB83.BRAB. Report on BW Policy (3 May 1950) Appendix. Extract from 1949 report on Biological Warfare BW(49)(40)(Final).

17 TNA, WO188/664. 100/BW/4/37. BW Sub-Committee Meeting (8 May 1950).

18 TNA, WO 188/664. BW(50)9 BW Sub-Committee. Biological Warfare Policy. Memorandum by the Chairman, BW Sub-Committee. (30 May 1950).

19 TNA, WO188/664. 100/BW/4/37. BW Sub-Committee Meeting (8 May 1950); WO188/668. AC10930/BRBM17, BRAB 17th Meeting (20 April 1950). Financial priorities fell outside of the ambit of BRAB, so it is likely that Crawford would not have consulted with them at all had economics been the only consideration.

The DRPC formulated their request soon after the explosion of the first Soviet atomic bomb in August 1949. This event had severely shaken Western assumptions about the state of Soviet defence research (Gowing 1974, Goodman 2003) and this unease probably lay behind the changing attitudes of the DRPC. In addition, the DRPC who had originally been excluded from decision-making over atomic warfare had begun to be given some influence in that domain (Agar and Balmer 1998). After the end of 1949 the DRPC were given authority to formulate policy on atomic matters as they related to non-atomic R&D projects.[20] This new and promising responsibility may further explain the turn from their initial championing of chemical and biological warfare.

The BRAB members' discussion and answers to the DRPC's questions further demonstrate how uncertainties could be drawn on as strategic resources by the expert advisors. Fildes, in this manner, asserted that 'years of fundamental research were still required to prove the possibilities of BW' and that it was necessary to be prepared for offence in order to deter an enemy.[21] Likewise, Henderson argued that although untried, the potential of biological warfare as a strategic weapon was great and that the UK was particularly vulnerable to a putative attack. Several other Board members added their view that germ warfare should be seen as both cheaper than and complementary, rather than equivalent, to atomic warfare. So, although the Board openly acknowledged the uncertain potential of biological warfare, this was used once again – and in contrast to the DRPC's interpretation of the implications of uncertainty – as an argument in favour of continuing research. In a report that followed from this discussion the Board declared in a somewhat bewildered tone that:

> in their view there was at present no justification for looking at BW as other than a form of warfare which might, by further research, be shown to be either practicable or, on the other hand, impracticable... on the balance of evidence, the development of a dangerous weapon was not improbable and certainly not impossible, *but only if active research, including field trials, was continued along the lines which had already been approved.*[22]

Despite this irresolute assessment, once again the scientists finished with a clear promise to resolve the uncertainty through further work. BRAB's views were duly passed on to the BW Sub-committee in writing and verbally by David Henderson.

---

20 Their ability to influence atomic matters was further enhanced by the appointment to the chair in 1952 of John Cockcroft, director of the Atomic Energy Research Establishment, Harwell. Full responsibility for atomic weapon R&D policy was not granted to the committee until 1954.

21 TNA, WO188/668. AC10930/BRBM17. BRAB 17th Meeting (20 April 1950).

22 TNA, WO188/668. AC10911/BRB83. Report on BW Policy (4 May 1950). Emphasis added.

The BW-Subcommittee's eventual submission to the DRPC in May 1950 followed an equally commendatory route, openly in favour of maintaining the existing programme. In the chair's memo, Crawford stated that although 'scientific opinion cannot guarantee that the development of a BW weapon is practicable', the available evidence certainly pointed towards success. Moreover, on the basis of best estimates biological warfare was 'of the same order of magnitude as the atomic bomb' and could be used to complement atomic weapons.[23] While he noted that uncertainty surrounding the potential of biological warfare could be construed as a reason for ceasing offensive research, he added that because intelligence information on Russia was 'extremely meagre… we are on the firmest ground if we base our appreciation on deductions from our own work'.

In June 1950, the Chiefs of Staff met to consider Crawford's findings. They endorsed the biological warfare programme and affirmed existing policy in the UK as being prepared for retaliation in kind. However, this was but a temporary reprieve. Over the next few years the status of biological warfare was to change quite dramatically. While policy statements during this period were ambiguous and uncertain, the overall direction of change was towards a defensive posture. In 1952, the Chiefs of Staff global strategy paper marked an overall shift in defence policy towards reliance on a nuclear deterrent.[24] Then, early in 1953, the Minister of Supply issued a directive emphasising defensive biological warfare research and suggesting that offensive research be confined to what it termed 'long range possibilities'. A defensive policy regime implied a delimited research programme, restricted in terms of resources, status and autonomy for both the researchers and their advisors.

In this unfavourable policy climate, uncertainties which had supported the expansion of the biological warfare programme were now turned against the research. To repeat Wynne's general comment, uncertainty became a problem only when interpreted as such. In a DRPC review of defence R&D in March 1954 developments in weapons of mass destruction were reported to have had fundamental effects on the defence programme only with regard to atomic weapons.[25] The review continued:

> It is still too soon in so new a field as biological warfare to forecast with any certainty what developments may yet arise... Development of offensive

---

23 TNA, DEFE 10/26. DRP(50)53. DRPC. BW Policy. Note by the Chairman, BW Sub-Committee (11 May 1950).

24 Followed in July 1954 by a Government decision to obtain a thermonuclear bomb (Arnold 2001).

25 TNA, DEFE 10/33. DRPC. Review of Defence R&D (10 March 1954). The review was largely dismissed by the Chiefs of Staff for reasons which have little to do with chemical and biological warfare and which are more concerned with the 'long haul' – a long term strategy which replaced the earlier concept of preparedness by 1957 – and the arrival of the hydrogen bomb. See TNA, AVIA 54/1749. MoD. DRPC Review (1954).

> biological weapons has been largely disappointing. We still know too little of the behaviour of agents under operational conditions to make the detailed quantitative assessments which could allow a strategic offensive based on biological techniques to be developed. Apart altogether from the political issues which must be faced before biological warfare could be initiated, the use of such weapons must be limited even in the strategic role to attritional forms of warfare. [26]

With regard to the future, the review's prediction for the state of development by 1957 pointed out that:

> sufficient knowledge of the behaviour of organisms may have been acquired for the potentialities of biological warfare to be assessed but this is by no means certain. No attempt will have been made to develop offensive weapons in the UK but the research will have continued to find suitable agents for different roles and suitable forms of weapons for their delivery and distribution. [27]

In drafts of the DRPC report, biological and also chemical weapons were removed from the list of areas of fundamental research to be accorded high priority, and only added again after some discussion, largely concerning the potential of new lethal chemical warfare agents (McLeish and Balmer 2012). And, only a few months later in July 1954, the Air Staff cancelled their standing requirement for an anti-personnel biological bomb. A year later, an ambitious series of sea trials which had taken place off the coast of Scotland, in the Caribbean and in the Bahamas was cancelled. Then, the BW Sub-Committee was dissolved thus severing a crucial link between the scientific advisors on BRAB and the Chiefs of Staff. And the 1955 DRPC review of defence research reported that biological research was now 'mainly defensively aimed'.[28]

## Returning to Uncertainty

I have briefly outlined two phases in the post-Second World War history of biological warfare policy – an expansionist phase immediately after the war and a reining in of the programme as the UK moved over to a defensive policy in the early to mid-1950s. Within the secret policy-making arena for germ warfare, the military and civil servants were largely concerned over whether or not biological weapons were a worthwhile investment. In order to address this question, advice on the current state of the art in biological warfare research, together with expert

26 TNA, DEFE 10/33. DRPC. Review of Defence R&D (10 March 1954).

27 TNA, DEFE 10/33. DRPC. Review of Defence R&D (10 March 1954).

28 TNA, DEFE 10/34. DRP/P(55)50 (Final). DRPC. Review of Defence Research and Development (23 January 1956).

opinion on the potential destructiveness of the weapons were obviously important. In providing this advice, scientists did not hide uncertainties from view or use them in an entirely negative sense.

Scientific advisors during the expansionist period openly expressed uncertainties which, tactically at least, might have appeared to diminish their cognitive authority but which actually worked as a crucial rhetorical device for sustaining the momentum of the UK research programme. Boundary-work here was not being utilised in an aggressive sense, to construct a confrontational boundary between science and policy. In a sociable setting it can be used to construct the legitimacy of particular roles and identities (Gregory 2000). While the advisors were not always firm advocates for the eventual success of the biological weapons programme, they remained keen advocates for the benefits of continuing scientific research in an offensive direction. All of these confessional claims to uncertainty did not, however, stand alone but possessed features which appealed, in turn, to various certainties.

First, as with the cases pointed out by Shackley and Wynne, the uncertainties were presented as tractable. This involved an appeal to the providence of science, a certainty that research would eventually resolve current uncertainties over whether or not a bomb could be produced, even if it could not guarantee a bomb. Second, the credibility of the uncertainty claims rested on the legitimacy of the advisors to pronounce on uncertainty, to be able to state authoritatively that 'we do not know' and also to specify what could not be known, in particular the impossibility of extrapolating results to real war situations. Third, and somewhat less obviously, the scientists' formulation of uncertainty appealed to their certain status as neutral advisors. By claiming uncertainty over whether or not a potent biological weapon could be delivered, the scientists appeared to be signalling that they had no stake in the outcome of the research. And, by implication, that they possessed no vested interests in the political decision to maintain or expand the research programme. This allusion to a form of Mertonian disinterestedness would have pre-empted any cynical response from the Chiefs of Staff that might have followed had they promised a weapon in more certain terms. Overall, the defence mounted by the advisors depended on both certainty and uncertainty claims – not working against each other, but rather propping each other up.[29]

---

29 A competent discourse analyst would most likely identify these features in terms of 'interest management', 'stake' and 'category entitlements'. See Potter (1996) Chapters 4–5. My aim here is less to tease out the precise features which contribute to the 'facticity' of the advisors' claims and rather to note how that facticity depends here on the symbiotic combination of both certainty and uncertainty. It is possible to conjecture that the 'second order' certainty claims might, in turn, be propped up with further uncertainty claims. For example, a parallel uncertainty argument to that made by MacKenzie (1990: 378) about nuclear missiles could be formulated, that even in the aftermath of a war scientists might conceivably claim the use of a biological weapon was unrepresentative or had not generated sufficient data.

This said, the claims of the scientists were, by themselves, not always persuasive. Uncertainty, like certainty, was a contestable and socially contingent phenomenon. With the fate of biological warfare policy, the position of the DRPC as both audience and fickle supporter for germ warfare researchers was important. At a time when the scientists on the DRPC were denied access to atomic matters and regarded biological and chemical weapons as their alternative province, then the uncertainty claims of the scientists carried weight. After the identity of biological warfare as a weapon of mass destruction had been settled and its fate tied to that of nuclear weapons, uncertainty was deployed in a more familiar way to discredit the programme. Once the DRPC began to gain a foothold in decision-making over atomic matters, then the committee was in a position to withdraw its full-blooded support for biological warfare research (Agar and Balmer, 1998). As nuclear warfare advanced steadily up the priority list at the expense of biological warfare, the patchy achievements of the biological warfare programme were flagged by the DRPC to bolster an eventual shift to a defensive policy. The uncertainties which pervaded the research were now presented as certain proof that the programme could not deliver.

So, the point I have been arguing in this chapter is not that uncertainty exists in policy disputes – that is an over-familiar observation. Rather, it is to pick up on the observation that uncertainty is constructed and, as such, can be a useful tool for scientists to direct against themselves when securing access to support and resources. Uncertainty, rather than certainty, can be where credibility is most at stake. In the case of biological warfare, struggles over the cognitive authority of science might possibly be characterised in terms familiar to sociologists of science: simultaneously as struggles over legitimacy, credibility, access to resources and so on. But, the contest here was less about knowledge *per se* as both certainty and uncertainty were mutually combined in arguments for the continuation and curtailment of the research. The battle over the distribution of authority was not synonymous with a battle to distribute uncertainty, but rather was manifested as a battle to set the *tone* of the debate. Put another way, it could be read as a battle over the conditions for the possibilities for these experts to be certain or uncertain. As such, this was not just a struggle over who had the authority to pronounce on matters of certainty and fact – but also who had the power and authority to express legitimate doubt.

Chapter 6

# Secrecy, Transparency and Public Relations: Opening Up Porton Down in the 'Year of the Barricades'

On 25 April 1967, the BBC's leading science documentary programme, *Horizon,* covered the subject of chemical and biological warfare (CBW). Although the programme dealt mainly with the historical and international context of CBW, it touched specifically on the work carried out at Porton Down and included an interview with the director of the Microbiological Research Establishment (MRE), Dr CE Gordon Smith. In the course of the interview, Smith stressed that the role of MRE was to undertake defensive research, adding that if the Government changed its policy in this respect he envisaged many of his staff resigning. Equally, Smith was at pains to deny that MRE was an unduly secretive place:

> Smith: It's perfectly possible to come to the front door of this establishment on a No.9 bus from Salisbury. You won't be admitted unless we know who you are and what you have come for but we have about two thousand or more visitors a year and anyone who has got a good and reasonable reason for coming and talking to us or seeing the establishment can do so.
>
> Interviewer: But you have the reputation for being a fenced-in-place, having gates and what-have-you. Are there fences?
>
> Smith: There is a fence around us which is to keep cows out. We don't want cows walking about on our nice lawns any more than anyone else does. At various times people have tried to give us this image and have even gone to the lengths of lying down in the field with the cows to take a picture of this building through a barbed wire fence. This, of course, we resent because it gives a completely wrong impression of this establishment.[1]

This was by no means the first time that representatives of Porton had complained about the 'wrong impression' that people might have of its work. Earlier in the year, Smith had publicly stated that 'the British Government has no requirement to be able to use biological warfare offensively' and described the role of the MRE

1 TNA, DEFE 13/997. Report MD137 Transcription of BBC 'Horizon' - Chemical And Biological Warfare (25 April 1967).

'to assess the risk to the people of this country of attack by biological weapons and to try and find ways to defend them', this time he was speaking in an interview for 'The Silent War', made for the ITV's current affairs series *The World Tomorrow*.[2] Both programmes contained no challenge to the claims made by Smith. The Silent War focused instead on the emerging call by scientists internationally to ban CBW weapons and to bring all research in the area into the public domain; while the response of the Horizon narrator to Smith's cows on the lawn showed little sign of dissent:

> I suppose it's inevitable that these Porton laboratories should acquire a somewhat evil reputation – no organization concerned with biological warfare could avoid it. Yet there's nothing particularly sinister to be seen here… Only 10% of it [the results of research] comes under the ban of official secrecy. One's constantly told that they don't make biological weapons, but they are solely concerned with defence. They call it a kind of public health, which I suppose it is.[3]

Over the next eighteen months, Porton's relationships with not only the broadcast media, but across the spectrum of its public relations, would be transformed. Public protest over chemical and biological warfare research had been sporadic since the early 1960s, but intensified throughout 1968 (Hammond and Carter 2002: 211–35). Protest was coupled with what the authorities regarded as increasingly hostile press and television coverage. This chapter describes how secrecy at Porton Down was replaced with a measure of transparency, culminating in a series of open days at the MRE in October 1968.

Previous chapters have largely focused on how the establishments at Porton Down operated through disengagement with the public. We have considered how press releases operated to manage information and provide a highly filtered version of activities in biological and chemical warfare research. Chapter two described how negative press coverage early in the Cold War was largely tackled by individual approaches to senior scientists and an open day to which only a select group of scientists was invited. This chapter turns its attention to openness and transparency, and develops this theme in the context of a period where staff at the Ministry of Defence and Porton Down were forced into a far more engaged relationship with the public, particularly through increased media attention. This, of course, was not a transition from complete secrecy to complete transparency. I will sketch instead how secrecy becomes a form of 'stage management', with controls set over just what is and is not said. In doing so, I want to revisit a metaphor for secrecy that is fairly common and indeed which I have put to use earlier in this book, the idea of the sphere of secrecy.

---

2 TNA, DEFE 13/997. Report MD69 Transcription of ITV 'The World Tomorrow' – The Silent War (20 January 1967).

3 TNA, DEFE 13/997. Report MD137 Transcription of BBC 'Horizon' – Chemical And Biological Warfare (25 April 1967).

The sphere metaphor implies a clear division between those inside and those outside the secret, two worlds of knowing and not knowing. In my discussion of ignorance and uncertainty in the last chapter, I began to challenge this rather too neat notion of two distinctive worlds inside and outside the secret, by drawing attention to the uses of ignorance and uncertainty by expert advisors *within* the bounds of secrecy. Then additionally, from the outside, the secret may never be fully concealed. As geographer Trevor Paglen has pointed out, secrets will always advertise their own presence, for example through a blank spot on a map or a redacted document in an archival index, in the same way a bandage both hides a wound and signals its existence (Paglen 2009: 17). By focusing on what might be regarded as the first public relations campaign by Porton Down, albeit by no means their first dealings with the public, a further facet of secrecy is highlighted. In response to calls for transparency, officials in Whitehall released information and eventually held an open day. But this in turn conceals as much as, or more than, it reveals. Here, degrees of secrecy and transparency operate as tools of governance that are zoned (Barry 2006). Secrecy, in this respect, and as I will further demonstrate in this chapter, becomes more akin to a series of concentric spheres of secrecy, partial revelation, and deeper opacity.

## Accessing Porton Down

Some months after the Horizon broadcast, the BBC once again approached the Ministry of Defence for permission to film at Porton Down. Roy Mason, the Minister of Defence for Equipment, held responsibility for matters at Porton Down and was supportive of the BBC request, 'so as to encourage responsible public assessment'.[4] Mason nonetheless felt it was appropriate to seek the Prime Minister, Harold Wilson's approval on this sensitive matter as Wilson had 'preferred [us] in the past not to go out of our way to promote interest in Porton Down'.[5] At the same time, the request was also referred to the Secretary of State for Defence, Denis Healey, who initially wanted to wait until planned civilian related work, for example on foot and mouth disease or air pollution, was ready to show to the cameras.[6] A briefing to Healey suggested otherwise, pointing out that enough civil work to highlight was already taking place at Porton.[7] Other 'rather tiresome' matters, such as permissions given the previous year for ITV and BBC access, also

4 TNA, DEFE 13/997. Ministry of Defence, John Peters to Peter le Cheminant, 10 Downing Street (6 February 1968).

5 TNA, DEFE 13/997. Ministry of Defence, John Peters to Peter le Cheminant, 10 Downing Street (6 February 1968).

6 There had been a major epidemic of Foot and Mouth Disease in the UK starting in October 1967 (Woods 2004).

7 TNA, DEFE 13/997. From E. Broadbent to Secretary of State. Proposed BBC1-TV Programme on Porton (15 February 1968).

made the proposal difficult to refuse. Not least, that the Bernal Peace Library was planning a conference at the end of February on chemical and biological warfare. The Library, founded the previous year, described itself as 'an educational trust' assisting 'in the struggle for peace and toward ensuring the realisation of the fullest potentialities of science in building a world prosperous without war' (Burhop 1968). Officials in the Ministry of Defence were far more disparaging, describing it as a 'front organisation' for the communist party, and suggesting that it was 'always possible that BBC1 in a sulk might pay more attention to their arguments if they do not get the other side'.[8]

Admitting the BBC into Porton obviously did not allow them unfettered access. Transparency would be controlled, in line with a policy that recognised the release of information was a form of impression management:

> These two establishments at Porton have, or can be thought to have, such sinister implications that for several years it has been MOD policy to counteract this kind of impression by giving as much information about their work as been consistent with security. There has been no attempt at a blanket of security such as applies generally to the work of AWRE [Atomic Weapons Research Establishment].[9]

In this vein, Healey was reported to be 'most anxious that we should do all we can to ensure that the programme puts the whole subject in reasonable perspective and brings out the high proportion of the non-security work carried out at the establishments'.[10] One of the conditions for allowing the BBC access was that Healey would have his chance to preview the film and commentary, thus enabling him to 'seek to persuade the BBC to eliminate any unsuitable material, although it might not be possible to insist on this'.[11]

The media was not alone in having to make special arrangements to access Porton. In 1967, Harold Wilson had refused a proposal for an all-party delegation of MPs to visit the research establishments.[12] At the same time as the BBC made its request to film at the site, the matter of access for MPs resurfaced. A Parliamentary Select Committee on Science and Technology had been established to undertake a general review of defence research, and a question arose over whether or not they

---

8 TNA, DEFE 13/997. From E. Broadbent to Secretary of State. Proposed BBC1-TV Programme on Porton (15 February 1968).

9 TNA DEFE 13/997. Publicity for Microbiological Research Establishment, Porton. Note by the Ministry of Defence (27 February 1968).

10 TNA DEFE 13/997. E. Broadbent (PS/Secretary of State) to PS/Minister(E). Porton.(28 February 1968).

11 TNA DEFE 13/997. D. Gruffydd Jones (Cabinet Office) to Ewan Broadbent. Porton.(29 February 1968).

12 TNA, DEFE 13/997. Ministry of Defence, John Peters to Peter le Cheminant, 10 Downing Street (6 February 1968).

might be granted a visit to Porton.[13] Matters were further complicated because one member of the select committee, Labour MP Tam Dalyell, had tabled a number of Parliamentary Questions (PQs) seeking information about British chemical and biological warfare research. He had become concerned about CBW after a visit to Porton in the early 1960s, and was now advocating greater openness in this area. As Dalyell explained in a BBC Home Service interview:

> It's not that I take the line that Britain in any way or British Governments, are being wicked and evil about it, obscene through this kind of warfare might be, but it seems to me that technology in this field has now reached a situation where small countries have the know-how, with limited industrial capacity, to build up a micro-biological and chemical warfare potential and this, I think, really does alter in a new context, the whole outlook of, shall we say, secondary powers towards chemical and biological warfare.[14]

Dalyell's views rendered the already difficult question of what information to make available to the Select Committee even more problematic. To begin with, it seemed that the Select Committee would be trusted with some measure of access to secrets, Healey and Wilson met with members of the Defence and Overseas Policy Committee and decided that the Select Committee members could, if they so wished, be told about non-secret research and furthermore they 'should be given a general picture of the secret work' being undertaken.[15] Contradicting this decision, a briefing written for Healey's private secretary argued that any putative Select Committee visit to Porton should be treated the same as visits to any other defence research establishment. Ground rules for such visits had been formulated by the Ministerial Committee on Parliamentary Procedure, which suggested that an establishment had authority at its discretion to disclose information up to the category of Confidential. As such:

> Any departure from the consistent application of 'Confidential' as an upper limit e.g. by giving "a general picture of the secret work", would be bound to lead to grave difficulties and embarrassment for not only us, but for other Departments who may be the subject of one or other of these specialist Select Committee's attentions.[16]

---

13 Select committees are established by Parliament to look into particular issues, take evidence and issue reports.

14 TNA, DEFE 13/997. Transcript. BBC Home Service 'The World This Weekend', Concern Over Chemical Establishments. (9 June 1968).

15 TNA, DEFE 13/997. D. Gruffydd Jones (Cabinet Office) to Ewan Broadbent. Porton. (29 February 1968).

16 TNA, DEFE 13/997 ... AUS(R&D) to PS/S of S. Porton. (7 March 1968).

Healey relented a little, responding via his Assistant Private Secretary that a 'very general picture' of the 'main areas of effort' could still be given without disclosing secret information.[17]

The Select Committee's visit to Porton Down took place on 6 May 1968. After being shown around the two research establishments, the committee took evidence, which included classified evidence. Shortly afterwards, on 26th May, the *Observer* newspaper published an article revealing names of several university researchers working in collaboration with Porton.[18] It also revealed some of the classified information that had been given to the Select Committee: the annual cost of the two establishments, a reference to the chemical warfare agent production plant at Nancekuke in Cornwall, and – of particular sensitivity – that the UK had no stockpile of chemical weapons so that it would take quite an effort to develop workable weapons systems (Balmer 2010). The source of the leak was soon identified as Dalyell, who was subsequently investigated by the Parliamentary Committee for Privileges, the body responsible for investigating alleged contempt of parliamentary privilege (Galbraith 2000). Their decision, that Dalyell had breached parliamentary privilege, was accepted by a vote of 244 to 52 in the House of Commons, which meant that Dalyell had to be formally reprimanded:

> As Mr Dalyell came into the Chamber, the Speaker drew out a black tricorn hat worn on these rare occasions and placed it over his wig. As the house watched in embarrassed silence, Mr Dalyell rose and stood tense and still while Dr King clearly distressed at what he was having to do, read out the reprimand in a low voice.[19]

## Protest

The May scandal was preceded and succeeded by wider public activity focused on Porton Down. Student protest preceded the *Observer* leak, while a high profile protest outside the Porton establishment followed in its wake. Since the early 1960s, Porton had received increasing attention, not just as we have seen from Dalyell's parliamentary questions and the media, but also through protests by groups such as the Campaign for Nuclear Disarmament (CND) and the more radical Committee of 100, whose leadership advocated change through civil disobedience. Although Porton's official historians, Hammond and Carter (2002: 211–35), place the earliest protests outside Porton's gates by peace campaigners as early as spring 1953, the first of a rising wave of protests took place in June

17 TNA, DEFE 13/997. From RM Hastie-Smith (APS Secretary of State) to AUS(R&D). (9 March 1968).

18 L. Marks and J. Slaughter, 'Porton Names Germ War Scientists', The Observer (26 May 1968).

19 Noyes, Hugh (1968), 'Speaker's reprimand for Dalyell', The Times (25 July 1968).

1963, organised by the Committee of 100. Porton had also been placed in the media spotlight in 1962, following the death from plague of one of the MRD staff, Geoffrey Bacon, who had been undertaking laboratory work on a plague vaccine (Carter 2000, Harris and Paxman 1982). By the late 1960s, protest against and public questioning of CW work were related particularly to the Vietnam war and the role of UK chemical weapons scientists in developing CS tear gas, which had been used in Vietnam by US forces.

Protest specifically against chemical and biological weapons research took place far more widely than outside the gates of the Porton establishments, whether bound up with Vietnam, nuclear disarmament or even antivivisection. Public protests against the Vietnam war were frequent in London, and in March 1968 anti-war protest spilled over from Trafalgar Square into violence at the US Embassy in Grovesnor Square. Significant student protests in the UK also took anti-chemical and biological warfare under their banner. Essex university students, as part of a wider protest against defence research on campus, successfully disrupted a visit by Dr T. Inch of the CDEE on 7th May. Inch had accepted a visit by the university Chemical Society to talk about the causes of chemical toxicity and how to treat it.[20] According to the *Times* newspaper, police with dogs were sent for when around 150 students disrupted the meeting, read out a statement condemning chemical warfare and demanded that Inch answer questions about chemical warfare research.[21] Other students sat outside in the corridor, preventing the security team from escorting Inch from the premises. Drawing a parallel with the student demonstrations in Paris, which had flared a day earlier on May 6th, the *New Statesman* magazine described this as the 'putsch at Essex' and detailed the hostile questioning of Inch:

> When asked why Britain was allowing Porton's own patented CS gas, which is extremely toxic, to be made under licence by the US, which is using it in Vietnam – it was also used by the Paris police against the students – Dr Inch is reported to have replied 'Because we need the money'.[22]

This account prompted a swift letter from the Ministry of Defence, claiming that Inch had been unable to deliver his lecture, so could not have responded as reported to the question. The letter also denied that CS was highly toxic or that the Government or anyone in UK sold 'poison and nerve gases or worse'.[23] Back at Essex the situation escalated after three students held to be responsible for the disruption were suspended until June. Around 250 students responded by

20 TNA, DEFE 13/997. TCG James (Chief of Public Relations) to PS/Secretary of State. Porton – Letter to New Statesman (29 May 1968).

21 'Police dogs went to Essex University', *The Times* (08 May 1968).

22 *New Statesman* – 17 March 1968 – London Diary (Tom Baistow).

23 TNA, DEFE 13/997. TCG James (Chief of Public Relations) to PS/Secretary of State. Porton – Letter to New Statesman (29 May 1968).

marching on the vice-chancellor's home in protest.[24] Forty staff members signed a petition threatening resignation if the students were not immediately allowed back to study, provoking a letter to the *Times* from a member of the Economics faculty who complained that the students had violated the idea of the university as a place of 'reasoned conversation and dialogue' rather than 'a guerilla training school'.[25]

With the Inch and Dalyell affairs still fresh, the Ministry of Defence staff now braced themselves for a much publicised Whitsun demonstration, planned by a variety of peace groups to take place in the first week of June, outside the Porton establishments. Christian CND, together with CND, the Anti-Chemical and Biological Warfare Group, and the student-led Southampton Peace Action Committee were all expected to attend. The Anti-Chemical and Biological Warfare Group was a recently formed coalition of existing organisations including CND, the National Peace Council, the British Humanist Association, the Fellowship of Reconciliation, and the Women's International League, whereas the more recently established Southampton Peace Action Committee was described by the Ministry of Defence as 'militant'.[26] Added to this, the *Times* warned that the demonstration would also be attended by 'anarchists'.[27] With some 300 demonstrators expected for a three day vigil, army representatives warned Healey's private secretary that, although Christian CND was not expected to be disruptive, 'civil disobedience cannot therefore be discounted'.[28] As a precaution, one regiment with some light armoured vehicles, helicopters, and military police in attendance were assigned to cooperate with the Wiltshire Constabulary and the Army Department Constabulary. The troops would be located inside the establishments 'but they will be kept out of sight as far as possible'.[29]

A different response to the spate of publicity surrounding chemical and biological warfare was for the Ministry of Defence to allow for more publicity, albeit publicity controlled by the authorities. Requests were granted to provide the *Times*, *Guardian* and *British Medical Journal* with help in writing articles about Porton. So, on the eve of the Whitsun vigil, the *Times* ran an article on the protest which aired the Porton perspective. Neville Gadsby and Gordon Smith, the directors of the Porton institutes, were both quoted, justifying their work as being

---

24 'Protest after three are rusticated: 250 in university march', *The Times* (11 May 1968).

25 'Rusticated students fight on', *The Times* 13 May 1968; Lloyd, Michael 'Clash with Authority at Essex', *The Times* (18 May 1968).

26 TNA, DEFE 13/997. Letter to PS/S of S from Lt-Col [signature illegible] MA to CGS (29 May 1968).

27 Wright, Pearce 'Army ring Porton against germ war protest', The Times (1 June 1968).

28 TNA, DEFE 13/997. Letter to PS/S of S from Lt-Col [signature illegible] MA to CGS (29 May 1968).

29 TNA, DEFE 13/997. Letter to PS/S of S from Lt-Col [signature illegible] MA to CGS (29 May 1968).

both necessary for defence and largely published in the open literature. Gadsby added 'I am proud of it, and that is not just sloppy sentimentality'.[30]

The Whitsun protest passed without violence:

> Led by Father Simon Blake, of St Dominic's Priory, London, chairman of the Christian group of CND, they carried out a ceremonial cleansing by throwing liquid disinfectant over the gates and the fences and spraying the air… To end their protest they joined hands, and played 'Ring-a-ring o' roses', and when it came to 'all fall down' they fell on their faces on the grass verges, lay there for about a minute, and vowed they would continue their opposition until they died.[31]

Within the Ministry of Defence and Porton Down, the unwanted spotlight on biological and chemical warfare was starting to create anxiety. This anxiety was further animated by talk within Whitehall and beyond of transferring the MRE to a civil authority (Balmer 2006). Dalyell, as part of his general efforts to penetrate the secrecy at Porton, had persuaded several eminent scientists to write directly to Harold Wilson at the end of May to request that the MRE be transferred from the Ministry of Defence.[32] The mood within Porton was captured in a letter to the Chief of the General Staff from the Master General of the Ordnance (MGO), who reported that on a routine visit to the MRE:

> The Director, in whose judgment I have every confidence, said that the tendentious and unbalanced presentation of so-called 'Germ Warfare' by press, television and radio was beginning to have adverse effect on the morale of the establishment. Although his scientists fully appreciated that the topic was controversial they felt aggrieved that the case for a defensive capability was seldom made publicly, nor were the myths about their activities contradicted officially. They would like some authoritative assurance that HMG [Her Majesty's Government] was serious in requiring this defensive capability to be maintained, particularly since the whole future of the establishment was known to be under consideration from the financial point of view.[33]

---

30 Wright, Pearce 'Army ring Porton against germ war protest' (1 June 1968).

31 'Ring o'roses ends CND protest', The Times (4 June 1968). The children's song 'ring-a-ring o' roses, a pocket full of posies, a-tissue, a-tissue we all fall down' involves joining hands, singing the song and falling down at the appropriate point.

32 Twenty one scientists had written to the Prime Minister by the end of June, all Fellows of the Royal Society, including eight Nobel prize winners. The scientists were: Lord Boyd-Orr; Lord Jackson of Burnley; Sir James Gray; Sir Francis Knowles; Professors M Abercrombie; C Auerbach; EHS Burhop; W Hayes; Dorothy C Hodgkin; H L Kornberg; R Markham; G. Porter; C F Powell; C H Waddington; Maurice HF Wilkins; Drs JH Humphrey; Max F Perutz; NW Pirie; Frederick Sanger; RLM Synge; Ms DMM Needham. MRE remained under military control throughout the 1960s.

33 TNA, DEFE 13/997. MGO to CGS. Biological Warfare (n.d.) [in file with papers dated end of May 1968]. The Master General of the Ordnance was responsible for both the

Consequently, the MGO and the Chief Scientist for the Army met and proposed that Healey pay an overdue visit to the Porton establishments.

### *A Plague on Your Children*

The BBC1 programme that had been filmed earlier in the year was due to be broadcast in the week following the Whitsun demonstrations. It was scheduled to be followed by a live discussion later the same evening and the BBC had requested that someone from the Ministry of Defence, preferably a scientist, appear on the programme. The Chief of Public Relations at the Ministry, Cecil James, predicted that this would be a 'fairly controversial' programme and warned that 'one must assume that any panel will include individuals who are violently opposed to work in the CBW field'.[34] Suspicion was cast not only on the external participants in a discussion panel; James further warned that anyone from the BBC was 'likely to be unsympathetic. So anyone who speaks in defence of the work at Porton is bound to have a hard time'. Gordon Smith was suggested as the most experienced person to face interrogation, and the Chief Scientist at the Ministry of Defence, zoologist Solly Zuckerman was also touted as a 'long shot', had he not been away on holiday at the time.

In the event, the BBC1 programme 'A Plague on Your Children' was broadcast at five past nine in the evening on June 6th without immediately generating too much fresh controversy for Porton. Indeed, the only fresh upset occurred behind the scenes as horrified Ministry of Defence officials watched the recently retired Director of CDEE, Dr Eric Haddon, admitting on camera the same 'secret' that had got Dalyell hauled before the Privileges Committee, that the UK had no stockpile of chemical weapons. The interview had taken place in March, but the BBC had not allowed the Ministry of Defence officials to see the programme beforehand. This revelation effectively cut short any attempt to claim the moral high ground by suggesting that the *Observer* or Dalyell had recklessly leaked classified information.[35]

Four days later the producer of 'A Plague on Your Children', Adrian Malone, was interviewed by Michael Dean on the BBC2 arts programme 'Late Night Line-Up'. Dean claimed that 'reaction to the programme from the general public has been overwhelming horror and revulsion', and speculated with Malone as to why he had been granted access in the first place:

---

Porton research establishments because they were Army Department establishments.

34 TNA, DEFE 13/997. From TCG James to PS/Secretary of State. BBC1 Programme about CBW (30 May 1968).

35 TNA, DEFE 11/672. Select Committee on Science & Technology – Porton Evidence (11 June 1968).

> Dean: Did any of the officials involved, especially those who gave you the final permission, ever express the view that the public would be fully in favour of Porton Down once this programme had been seen?
>
> Malone: Yes I think there was… I think in general they expected to see that when the research was put on the screen, that people would see the value of it.
>
> Dean: They have been proved somewhat wrong
>
> Malone: I think so.[36]

Dean next suggested that the programme could be criticised because it had contained no overt discussion of ethics, to which Malone replied that he just wanted to present the facts and start a public discussion:

> Dean: In other words, you set out to educate people, but you also realised that you'd be scaring the pants off them?
>
> Malone: I don't think you can avoid scaring the pants off people when you talk about CBW.[37]

In the days that followed a 'Plague on Your Children', the Ministry of Defence appeared to be more optimistic that the BBC had avoided scaring anyone's pants off. Cecil James, the Chief of Public Relations at the Ministry of Defence, reported that recent access granted to journalists from the *Times*, *Guardian* and *Telegraph* had resulted in coverage that was 'as we would have wished'.[38] The *Times*, for example had run a supportive editorial entitled 'Defence Against Germs' on the same day as the BBC1 programme. It compared Porton Down with Aldermaston, the headquarters of the British atomic weapons programme, and the site of annual protest marches in the late 1950s and early 1960s:

> Porton Down is now displacing Aldermaston in the demonology of collective protest against the horrors of war. On the plane of emotion the switch is appropriate. The potentialities of chemical and microbiological warfare are as ghastly as anything in the nuclear nightmare. On the plane of policy the protest is even less helpful than the unilateralism of the Aldermaston marches in their

---

36 TNA, DEFE 13/997. Transcript: BBC2 'Late Night Line-Up'. Work at Porton Down (10 June 1968).

37 TNA, DEFE 13/997. Transcript: BBC2 'Late Night Line-Up'. Work at Porton Down (10 June 1968).

38 TNA, DEFE 13/997. PR Policy for Porton. TCG James, Chief of Public Relations (10 June 1968)

> heyday. In the present state of the world the work carried out at Porton Down is a source of reassurance, not a cause of panic.[39]

The Ministry itself was starting to organise itself in response to public criticism and protest. James had recently met with representatives from the Porton establishments to discuss public relations, and reported their gratitude at some loosening of the 'rigid central control' that had prevented heads of the establishments from responding 'factually' to the press, although they still avoided statements about policy. The meeting had also generated the suggestion that Porton hold some sort of open day for the wider public so that, in James' pragmatic terms: 'We may never succeed in making Porton popular but we should try to disperse as far as possible the aura of secretiveness which so many people think hangs around this kind of work'.[40]

Prime Ministerial approval for open days followed swiftly, while inside the Ministry of Defence a beleaguered atmosphere remained as the new Minister of Defence for Equipment, John Morris, met with colleagues to 'review what we were doing to counter the present campaign of criticism against the establishments'.[41] Largely recapping the measures outlined by the Chief of Public Relations, Morris also considered the Nancekuke plant in Cornwall, 'which had become the target for special criticism'. The CND magazine, *Sanity*, had just published an aerial photograph of the site with the caption 'The picture no one dare print' and this in turn had made national news.[42] In response, Morris and his colleagues suggested that Nancekuke could be opened to journalistic scrutiny or, in a move that equated responsible journalism with state controlled journalism, to allow 'at least a limited facility to a responsible journalist, for instance to Westward TV [the regional TV station] which had asked for this and had offered to allow the script to be vetted'.[43] Tellingly, the main security risk at Nancekuke was 'the need to conceal how limited our resources were' and this lack of capability was deemed a serious enough risk for the Minister to decide not to permit any media access to the site.[44]

*Points of View*

Any hopes that the storms around Porton had abated were dashed when, one week on from 'A Plague on Your Children', feedback on this programme was aired

---

39 'Defence Against Germs', *The Times* (6 June 1968).

40 TNA, DEFE 13/997. PR Policy for Porton. TCG James, Chief of Public Relations (10 June 1968).

41 TNA, DEFE 13/997. Porton – Publicity – Note of a Meeting held by Minister (E) on Wednesday 12 June (12 June 1968).

42 'Ministry of Defence to check CND pictures', *The Times* (8 June 1968).

43 TNA, DEFE 13/997. Porton – Publicity – Note of a Meeting held by Minister (E) on Wednesday 12 June (12 June 1968).

44 TNA, DEFE 13/997. Porton – Publicity – Note of a Meeting held by Minister (E) on Wednesday 12 June (12 June 1968).

on the BBC's 'Points of View'. As suggested by the title, the format of 'Points of View' involved the presenter, Robert Robinson, introducing items with actors then reading out excerpts from letters sent to the BBC. In this instance, viewers' points of view were overwhelmingly negative. 'Shocking and horrifying' wrote one Miss G. Patterson of Edinburgh, while a John Swain from Swanage thought that the scientists were 'beyond the understanding of ordinary men'.[45] Others were appalled at the scenes of vivisection shown on their screens. While these opinions were hardly welcome to the Ministry of Defence officials and Porton scientists, what really provoked their anger was that Robinson betrayed any pretence to neutrality or deference for Porton. Introducing the item, Robinson had immediately made his views apparent:

> Something somewhere is awry when a bunch of intelligent University educated men spend their working lives devising plagues to kill people with, you've only got to say it to know. A programme called 'A Plague on Your Children' showed them doing it, down at Porton down. Bottling the bacteria as though it were so much stewed fruit.[46]

And when M. Bagshaw of West London complained that the BBC had 'presented the whole thing in an unfavourable light, you weighted it against Porton', Robinson retorted:

> Well now, just how do you present the whole thing favourably, I mean supposing you were a scientist and got home tired and happy at 5.30 and your wife said 'good day, dear?' and you said: 'oh yes, we made enough plague germs to wipe out ten million people'. Well now, how would you say that favourably?[47]

The programme, according to the Army Chief Scientist, RM Wilson, 'created a crisis situation at the MRE'.[48] Wilson moved swiftly and, in consultation with MRE's Gordon Smith, drafted a complaint to Lord Hill of Luton, the Chairman of the BBC Governors. The letter, quoted here at length to convey its furious tone, was intended to be sent in the name of the Secretary of State for Defence:

> I find this a disgraceful document; it is the most vicious, outrageous and unjustified attack on a group of public servants which I have yet experienced. Everything

45 TNA, DEFE 13/997. Verbatim Transcript of Item on BBC Television 'Points of View' (8.45pm, 12 June 1968).

46 TNA, DEFE 13/997. Verbatim Transcript of Item on BBC Television 'Points of View' (8.45pm, 12 June 1968).

47 TNA, DEFE 13/997. Verbatim Transcript of Item on BBC Television 'Points of View' (8.45pm, 12 June 1968).

48 TNA, DEFE 13/997. RM Wilson Chief Scientist (Army) to PS/Minister (E) (14 June 1968).

> which Mr Robinson said can be construed as a deliberate misrepresentation of the functions of the Microbiological Research Establishment and of the work on which its scientists are engaged… I am sure that if you study the script of this programme you will agree it constitutes so gross perversion of the facts as to be a grave abuse of responsibility under the Charter of the British Broadcasting Corporation… I can find no words strong enough to express my contempt for this cowardly attempt to subject able and dedicated scientists to public odium and disrupt the work which they have undertaken in the national interest.[49]

As the letter went through successive drafts, the outraged tone remained but with some tempering. A preliminary paragraph was added saying that 'A Plague on Your Children' was 'no cause for complaint', only the subsequent comments by Robinson.[50] A threat was initially made to release the official response to the press, but this later became an exhortation to redress the 'very serious misrepresentations of the facts', and the attack on 'certain public servants, known by name and with professional careers as well as private lives at stake, whose work is designed solely to help our community' in a future Points of View programme.[51]

Undoubtedly the impetus for toning down of the complaint came from John Morris' private office. In a letter to Healey's private secretary, John Peters, Morris' private secretary, explained at length the need for some measure of diplomacy. According to Morris, he wrote, there were different categories of critic:

> The reasonable and responsible ones can be reached by patient repetition, in Parliament and outside, of the MOD's point of view. For example, the decision to hold open days at Porton, even though these will be by invitation and restricted, has taken a good deal of heat out of this 'moderate' criticism… Then there are the rational critics whose deliberate aim is to damage the defence effort, and the irrational ones who follow the latest protest fashion, who are equally beyond the reach of argument and must be ignored, and endured. Finally, there are those who should know better, but deliberately overstep the bonds [sic] of acceptable comment, and Mr Robinson's performance of 'Points of View' is a particularly bad case in point.[52]

Peters proceeded to suggest that, while it was necessary to set the record straight, the 'respectable critics' needed to be kept under control. The original drafts of the letter to the BBC were worded so strongly that they might 'provoke a reaction in favour of the wretched Mr Robinson – who is after all a not very important

49 TNA, DEFE 13/997. Draft Letter from Secretary of State for Defence to The Chairman of the British Broadcasting Corporation (n.d.).

50 TNA, DEFE 13/997. Draft Letter from S of S to Lord Hill of Luton (n.d.).

51 TNA, DEFE 13/997. Draft Letter. (n.d.).

52 TNA, DEFE 13/997. From John Peters, Private Office (E) to PS/Secretary of State. Criticisms of Porton – BBC T/V Programme 'Points of View' (14 June 1968).

performer in a not very important programme – and draw attention to his views'.[53] If such a strong letter was made public and provoked a wider sympathetic response, Peters concluded, Lord Hill might feel obliged to defend Robinson.

This position was also supported by John Groves, the new Chief of Public Relations, who reiterated that to date only a relatively small audience had watched Robinson on 'Points of View' and that no risk should be taken to have his remarks more widely circulated. There did also appear to be a period of brief reprieve for the Ministry of Defence and Healey was reported to believe the topic was 'perhaps going off the boil'.[54] As part of the fallout from the Essex university protests, students had organised a demonstration in London for 16th June. This was regarded by the Ministry of Defence as 'a flop'.[55] While the day after, the *Guardian* headline ran 'Skeleton Turn-out for Germ Warfare "March of 2000"'and the *Times* ran the dismissive headline: 'Only 200 in germ war demonstration', then described how the protesters had marched from Hyde Park to Whitehall, handed in a petition, and chanted 'close down, Porton Down'. The article also reported on a scuffle with the Horse Guards as some protesters 'forced their way into one of the two sentry boxes where a mounted trooper is on day-long motionless guard. The horse in the box, called Shannon, became distressed and restive at the commotion'.[56] Although swords were drawn, there was no blood shed and the crowd dispersed.

Quite possibly buoyed up by the feeling that the Ministry of Defence was beginning to turn the tide of the public relations battle, Groves took it upon himself to write directly to Robert Robinson, arguing that the role of Porton was to protect troops and civilians from chemical and biological weapons, and finishing with a parody of Robinson's own words:

> Something is awry when an intelligent university educated man allows unreasoning emotion to betray him into turning the facts on their head and committing a grave injustice against those whose lives are devoted to saving our children from the plague.[57]

Unrepentant, Robinson wrote a few days later to Groves that, because of secrecy, there was no way of telling whether or not staff at Porton were working for defensive or offensive purposes. Then he added that, even if the work was to aid defence, this still did not preclude use of biological weapons: 'But once the

---

53 TNA, DEFE 13/997. From John Peters, Private Office (E) to PS/Secretary of State. Criticisms of Porton – BBC T/V Programme 'Points of View' (14 June 1968).

54 TNA, DEFE 13/997. From E. Broadbent (PS/Secretary of State) to CPR (18 June 1968).

55 TNA, DEFE 13/997. From E. Broadbent (PS/Secretary of State) to CPR (18 June 1968).

56 'Only 200 in germ war demonstration', *The Times* (17 June 1968).

57 TNA, DEFE 13/997. JD Groves to R. Robinson, BBC Television (17 June 1968).

principal [sic] of manufacturing germs for defence is accepted, the possibility of their use (in a defence situation) is also accepted. On this ground alone, I feel that my own words – which I hope were both forceful and temperate – were justified'.[58]

In the few days between Groves' letter and Robinson's response, the BBC had again broadcast an attack on the work at Porton. 'Twenty-four Hours' was a current affairs programme that aimed to break the mould of deferential news coverage by adopting an investigative and confrontational style of reporting. Biological and chemical warfare was covered by the Twenty-Four Hours team for some twenty minutes on the evening of 18th June. Presenter, Kenneth Allsop, introduced an interview with biochemist Steven Rose, who had been one of the organisers of the Bernal Peace Library conference, and with CDEE director Neville Gadsby, by pointing out that the main objection by scientists and MPs was to the secrecy surrounding work at the establishments. This point was reinforced by Rose, who clarified that he was:

> against the development of chemical and biological warfare as such, what I'm not against is the development of genuine defensive measures against a potential attack by chemical and biological agents. So far as the open days at Porton are concerned, what worries me is not the open day but what goes on the other 364 days and it doesn't seem to me that this is going to change the situation at all.[59]

To which Allsop added:

> Well, perhaps I'm in the minority but the entire enterprise of perfecting ways of killing people by deliberately spreading germs and chemicals seems unutterably loathsome and the idea of winning the hearts and minds of people by having jolly open days, seems to me to be a diseased idea in itself, or do you think I'm wrong?[60]

Gadsby was then allowed to defend the establishments:

> I would like to introduce a sense of realism about this claim that Porton is a super-secret establishment. It is far from being a super-secret establishment and this concept has been built up by the public opinion techniques. The establishment,

58 TNA, DEFE 13/997. Robert Robinson (BBC) to JD Groves, Chief of Public Relations, Ministry of Defence (21 June 1968).

59 TNA, DEFE 13/997. Transcript of BBC "Twenty Four Hours", Research into Germ Warfare. (18 June 1968).

60 TNA, DEFE 13/997. Transcript of BBC "Twenty Four Hours", Research into Germ Warfare. (18 June 1968).

> like all defence establishments, must have an element of its programme which is classified and this is characteristic of all defence research and development.[61]

While this was certainly more credible than Gordon Smith's argument just over a year before that Porton's fences were to keep the cows off the lawns, unlike in that interview there was now little sense that the scientist's position was beyond reproach. Indeed, Allsop finished the interview with the words 'well, gentlemen, while you go on producing bubonic plague and nerve gases at Porton that's as far as we can go here. Thank you very much'.[62]

*Counterattack*

The observation just days before that this topic had 'gone off the boil' now looked hopelessly flawed. Healey agreed to visit both Porton establishments in the coming month, and fired off a letter of support to Gordon Smith at the MRE, which is worth quoting at length to show how the Secretary for Defence expressed just how badly he felt the scientists had been treated, but also laid out the Government position and tried to finish on an optimistic note:

> I should like you to know how much I sympathize with you and your staff over the disgraceful campaign of vilification, to which you have been subjected. I would like to stress that I and my colleagues in the Government are fully aware of the facts in the current public controversy over Porton. We recognize that there is a genuine threat in the field of chemical and biological warfare, that your work is solely defensive, and that it has a worthwhile 'spin-off' in terms of hardware and the advance of scientific knowledge. We have tried to bring all this home by means of statements in Parliament and by giving the facts to the press; and I think we are now beginning to have some success. Your own personal appearances on television have played an important and constructive part. The protest here last weekend was as you know, a complete flop; that dismal demonstration was reduced to taunting a couple of soldiers.[63]

Next, the letter planned for Lord Hill of Luton was sent, complaining about both 'Points of View' and 'Twenty-Four Hours'. Although strongly worded, Healey was keen not to appear to advocate censorship on sensitive matters, but instead took the line that factual errors had been made, a right to reply had been quashed, that he had a duty to protect staff undertaking a difficult task, and finally that he was urging the BBC to maintain its reputation for impartiality. The tone of the

61 TNA, DEFE 13/997. Transcript of BBC "Twenty Four Hours", Research into Germ Warfare. (18 June 1968).

62 TNA, DEFE 13/997. Transcript of BBC "Twenty Four Hours", Research into Germ Warfare. (18 June 1968).

63 TNA, DEFE 13/997. Denis Healey to Dr CE Gordon-Smith (21 June 1968).

letter and its language of governmental facts versus journalistic ignorance are best gauged, once again, by quoting it at length:

> From time to time my Department and I have had differences of opinion with the BBC. Since I have a high regard for the standard of impartiality which the Corporation usually achieves, and I also value your right of free comment, I do not raise these difficulties with you. But the recent TV treatment by the BBC of the work being done at MRE Porton is something which I feel I must not let pass, without first drawing it to your attention. Not only have certain members of your staff revealed a dangerous ignorance of this subject; they have also displayed an unwillingness to consider the true facts of the case, which has even gone so far in one instance as to deny us the usual right to reply… In writing to you in these terms I have, of course, got a departmental axe to grind. But I am also very conscious of my responsibility to the staff at Porton who do a difficult and sometimes dangerous job and deserve a good deal more public appreciation than they are at present getting. Finally, I am sincerely concerned for the BBC, whose normally high standards seem to be slipping on this occasion.[64]

John Morris, in his capacity as Minister of Defence for Equipment, also took a more proactive stance on publicity, agreeing to appear on Welsh Harlech TV because it provided 'an opportunity to reply to a good deal of misinformed or hostile criticism'.[65] The questions were provided in advance and gave Morris' private office ample opportunity to draft replies. The interview briefly covered the work at Porton, moved on to the use of CS gas in Vietnam, with Morris arguing that the US authorities would not need to inform the UK of its use in Vietnam, but that it was untrue that it had caused deaths; finally Morris was given the opportunity to defend the work and workers at Porton. Hand scribbled on the cover note to the draft answers is a wry note addressed to Healey stating: 'Nothing new here, but I doubt that it has been said in Welsh before'.[66]

Plans for the open day at Porton had started to gather pace and dates were proposed for the end of October. This was a far shorter period than the usual nine months that other establishments took to plan their open days, which meant that work was 'disrupted to some extent' as 'emergency measures' had been introduced.[67] A committee of scientists had been formed in order to design exhibits and demonstrations, other staff gathered information on open days that had been held at other defence establishments, and a preliminary guest list was drawn up.

64 TNA, DEFE 13/997. From Denis Healey to Lord Hill (25 June 1968).

65 TNA, DEFE 13/997. John Peters (Private Office (E)) to RJ Brazier, Chief Whips Office (24 June 1968).

66 TNA, DEFE 13/997. John Peters (Private Office (E)) to Head of MGO Secretariat (24 June 1968).

67 TNA, DEFE 13/997. To Minister (E) Open Days at Porton (21 June 1968).

This guest list was carefully circumscribed. The limitation of the list to include some and exclude others has much in parallel with Shapin's argument that displays of scientific proof at the early Royal Society operated through carefully engineered social, material and literary technologies (Shapin 1988), where only people of particular status were judged to be credible witnesses to knowledge production. The initial guest list for the Open Days proposed that some 500 people each day would visit, divided into a dignitaries day (MPs, senior local council officials, industrial representatives and other senior people), a senior scientists day, and a scientists' and laymens' [sic] day. The last day also included a small number of places to be 'allocated on bona fide application from the general public'.[68] The planners also suggested to John Morris, with evident distaste, 'we could also include on the last day representatives of National or local "peace" associations if you think this is on balance desirable'.[69] Morris responded that the local peace associations, 'whoever these may be', would not be welcome at the open day, but neither would relatives of Ministry of Defence staff, nor children and teachers from local schools.[70] Five hundred attendees, Morris further suggested, 'runs the risk of sacrificing instruction with congestion'.

Another issue that concerned those planning the open days was whether to address concerns about vivisection. While this issue had not yet attracted the full attention and protest of anti-vivisection groups, Hammond and Carter (2002: 229) note that around this time there was a 'slow trickle' of parliamentary questions about animal experimentation at the establishments. Each year Allington Farm on the Porton site bred approximately 600 cats, 10,000 rabbits, 30,000 guinea pigs, 50,000 rats and 100,000 mice for use in experiments.[71] In preparation for the Open Day, the question arose as to whether the farm would remain behind closed doors and one briefing for John Morris declared: 'There is little doubt that a visit to Allington farm would give some of the Press the time of their lives in reporting on the kittens and puppies which are destined for use in experiments, but perhaps it is best to face this once and for all'.[72]

The sensitivity of the issue was such that Denis Healey also became involved in the debate over whether or not to open up Allington Farm, with a note within the Ministry of Defence reporting that:

> S of S [Secretary of State] has expressed the hope that it may prove possible to concentrate the public gaze on the rats, mice and guinea pigs kept for experimental purposes. If puppies and kittens are allowed a prominent position in

68 TNA, DEFE 13/997. To Minister (E) Open Days at Porton (21 June 1968).

69 TNA, DEFE 13/997. To Minister (E) Open Days at Porton (21 June 1968).

70 TNA, DEFE 13/998. To MGO from Office of Minister (E).Porton – MRE Open Days (25 June 1968).

71 TNA, DEFE 13/998. Visit to Allington Farm, Porton, 20 February 1968. Brief for US of S (Army).

72 TNA, DEFE 13/997. To Minister (E) Open Days at Porton (21 June 1968).

> proceedings, an unfavourable public reception to the work of the establishment is guaranteed.[73]

Allington Farm remained closed to the public. Dates for the open days were eventually finalised for 23–25 October 1968. And the number of guests each day was reduced, with preference given to 'appropriate' dignitaries including MPs, the national and scientific press, broadcast media reporters, representatives of research associations and industry. While this did not completely exclude national scientific associations, senior government officials, or even a small number of the general public, the choice reflects the underlying aim of the open days to produce a positive image of the establishments to a select group of witnesses.

The general attempt to rally against the attacks on Porton improved the situation for chemical and biological warfare researchers. Healey received swift responses from Gadsby and Gordon Smith, thanking the Secretary of State for his letters of support, Gadsby elaborating:

> Taking everything into consideration the morale of the staff here is remarkably high. Nonetheless the continuous sniping (and worse) which we have experienced in recent weeks has been wearying and in this country area it tends at times to involve not only members of staff but also their families.[74]

The letters were followed a day later by a response from Lord Hill of Luton admitting that the Twenty-Four Hours episode had 'caused a great deal of concern at the BBC', which had fallen 'short of the standard of impartiality expected from it'.[75] Action, Lord Hill informed the Secretary of State, had been taken, although no detail was given about what this action was. Healey later received a second apology from Lord Hill of Luton on behalf of the BBC. In this letter, Lord Hill distanced himself from the 'Points of View' presenter, writing that Robert Robinson's 'statement was too emotional' and furthermore that Robinson's later response had been typed on BBC notepaper and so gave the impression that his reply represented the views of the BBC.[76]

The extent to which the 'continuous sniping' had affected morale at the establishment can be gauged through the reaction of the Institution of Professional Civil Servants. The events of the previous months spurred the chairman of the Army Department, Scientific Staffs Branch of the Institute to write directly to the Chief Army Scientist, HM Wilson, complaining that 'these scurrilous attacks are

---

73 TNA, DEFE 13/998. From RM Hastie-Smith APS Secretary of State to PS/ Minister (E). Open Days at Porton (25 June 1968).

74 TNA, DEFE 13/998. From Gadsby to Healey (S of S for Defence) (25 June 1968).

75 TNA, DEFE 13/998. Lord Hill of Luton to Denis Healey (26 June 1968).

76 TNA, DEFE 13/998. Letter from Lord Hill of Luton to Dennis Healey (9 July 1968).

all the more vicious because the Official Secrets Acts prevent staff from making personal representations to either the BBC or the press'. The letter implored the Government to 'protect and defend staff from malicious attack'.[77] Wilson responded with assurances that there had been an immediate representation to the BBC 'regarding the grossly offensive nature of the comments', adding a paragraph that echoed the classification of critics that had come from the Minister of Defence for Equipment's office days earlier:

> It would, of course, be too much to hope that we shall see an end of the prejudiced criticisms of Porton and its staff; there will always be irrational critics who are beyond the reach of any argument and who must be ignored and endured. Our aims are to reach the reasonable and responsible critics and so gradually build up a better climate of public opinion. We also intend to deal promptly with those who should know better but appear to over-step the boundaries of acceptable comment.[78]

While now appearing to be retaliating against the spate of bad publicity, officials within the Ministry of Defence were given little respite. On 27th June, Haddon, the retired director of CDEE passed on a letter he had received from a former RAMC (Royal Army Medical Corps) officer, Richard Adrian, warning that the *Observer* had attempted to get information from him about his time at Porton in the early 1950s. The reporter wanted information about Flight Lieutenant Cockayne, who had appeared on a BBC programme and attributed his current mental illness to his time at Porton. Adrian had agreed to be interviewed in person at his home in Cambridge, but only if accompanied by a witness. Moreover, Adrian added: 'I had also intended to tape record the interview, being inexperienced in such things, pressed the wrong knob and failed to record anything. Smith [the reporter], however is under the impression the interview has been taped'.[79]

A swift enquiry within the Ministry revealed that Cockayne, who had voluntarily retired in 1954, served in the Munitions Division at Porton. His job involved supervising the loading of bombs onto aircraft for experimental trials.[80] During his spell at Porton he had been treated twice, once for mild myosis (dilation of the pupils) possibly because of exposure to nerve gas, and secondly for an infected boil on his head. Shortly afterwards, more information came to light

---

77 TNA, DEFE 13/998. B. Sutton, Chairman, The Institution of Professional Civil Servants (Army Department, Scientific Staffs Branch) to Dr HM Wilson, Chief Scientist (Army) Ministry of Defence. BBC Documentary – A Plague on Your Children (27 June 1968).

78 TNA, DEFE 13/998. HM Wilson to Sutton (11 July 1968).

79 TNA, DEFE 13/998. From Richard H. Adrian to EE Haddon (27 June 1968).

80 TNA, DEFE 13/998. Draft Brief for the Press Office attached to Gordon Hay (Defence Press Office (Army) to CPR. Observer Enquiries – Porton. F/Lt WN Cockayne RAF (Retd) (3 July 1968).

that, according to the Ministry's Chief of Public Relations, John Groves, the Flight Lieutenant had 'a record of instability' and his 'troubles certainly began before 1953'.[81] At this point the *Observer* was 'fighting shy of the story', but the Ministry prepared a press brief for 'if we are pushed into replying'. It gave away minimal information but admitted that Cockayne was 'involved in a number of technical activities' and that his treatment for myosis 'could be consistent with an extremely small dose of nerve agents'.[82] Groves was also quite explicit about the strategy underlying this briefing, portraying Cockayne as an unreliable source and suggesting that the Ministry had investigated his personal circumstances:

> This man has a long record of instability in the RAF and this seems to have worsened since his retirement. He has had constant trouble with money, drink and women. He has a police record in Australia since his retirement. The Observer are obviously worried about the accuracy of the facts he has been giving them, and I think our objective should be to show that these facts are wrong wherever it is possible to do so, in order to head off any follow up stories next week.[83]

By mid-July the various responses by the Ministry of Defence to the media and public had coalesced into a more coherent strategy. Morris, in response to another request, this time by ITN to make a five minute film to use as background for news items, informed Healey that 'it is now our conscious policy to gradually build up a climate of informed opinion on Porton without being deterred by the irrational attacks we may have to bear in doing so'.[84] At Porton Down, Hill's letter had been circulated to staff and their morale was reported to be much higher, although they still held some reservations about allowing further exposure to the media.

Healey's visit to Porton took place on 16 July and was certainly not made in secret, the press were alerted and the following day Healey announced in a Commons debate on CS gas that 'I had an opportunity to sniff CS gas at Porton yesterday, and I am glad to say that I am here today'.[85] During the visit both of the directors, together with representatives from the Whitley Council – the nearest equivalent to a civil service trade union – took the opportunity to express their

81 TNA, DEFE 13/998. John Groves (CPR) to PS/Secretary of State (5 July 1968).

82 TNA, DEFE 13/998. Handwritten note stapled to Groves to S of S 'The Case of Fl.Lt Cockayne. 5 July 1968; Draft Brief for the Press Office attached to Gordon Hay (Defence Press Office (Army) to CPR. Observer Enquiries – Porton. F/Lt WN Cockayne RAF (Retd) (3 July 1968).

83 TNA, DEFE 13/998. The Case of Fl. Lt. Cockayne (JD Groves) (4 July 1968). The story eventually ran in August: 'RAF Man "Victim of Porton Nerve Gas" … ', The Observer (11 August 1968).

84 TNA, DEFE 13/998. JM to Secretary of State. Porton – request for ITN facility (15 July 1968).

85 *Hansard* HC Deb 17 July 1968 vol 768 cc1418-20.

concerns over the spate of bad publicity. According to Healey, they did not resent 'proper public debate on the major political, military and moral issues in B & CW but about the publication of gross factual errors and personal attacks'.[86] To allay their concerns, Healey requested that John Groves and John Morris take stock of how they were waging their public relations campaign. And indeed, several changes had already taken place to speed up responses to any media comment. One press officer in the Ministry of Defence press room had been assigned special responsibility for Porton Down; the Porton directors were given a degree of freedom in dealing with local press enquiries without having to refer back to Whitehall. Morris, consequently, was able to report back to Healey that 'at least we are now free to deal with the publicity aspects of Porton as we see fit, and already I think there are good signs that our more helpful attitude is having a good effect on the attitude of rational commentators'.[87]

The open days eventually took place between 23rd and 25th October 1968 as planned, with a wide variety of attendees. The days paved the way for other events the following year, including a similar series of open days at Nancekuke, and a seven hour long 'teach-in' on chemical and biological warfare held at the University of Edinburgh. The 'teach-in' was attended by speakers from Porton Down and the Ministry of Defence, including John Morris as Minister of Defence for Equipment.

## Conclusion

This chapter has done anything but tell a story of a completely secretive organisation being forced into the open. Instead, what we see is a fairly rapid transition from secrecy that was, to a large extent, embedded in the institutional culture to an organisation that had to take steps to actively manage openness and secrecy. Previous press releases discussed in earlier chapters, and the television broadcasts in 1967 show that Porton representatives were not wholly unused to dealing with the media. But Minister of Defence Roy Mason's hope that approval of the 1968 request by the BBC to film 'A Plague on your Children' would 'encourage responsible public assessment' (rather, public approval) was derailed by rising protest against secrecy at the two establishments.

This transition took place in the context of more widespread demonstration and unrest, not just in the UK but globally, during the 'year of the barricades' (Caute 1988). Porton Down was but one focus, albeit a symbolically important site, of protest. As historian Jon Agar has pointed out, this was a period of change for science in the UK: previously closed discussions among experts on all manner of scientific topics were increasingly being subjected to public scrutiny, and relatively

86 TNA, DEFE 13/998. From Healey to Minister (E). Porton .MO/26/14.(17 July 1968).

87 TNA, DEFE 13/998. JM to Secretary of State. Porton. (22 July 1968).

accessible scientific establishments – compared, say, with the many desert sites used in the USA – were becoming sites for demonstration and protest (Agar 2008). Within this context, Porton Down was not singled out for harsh media treatment. Quantitative research using content analysis has shown how science coverage in the UK print media shifted quite dramatically in the late 1960s from a deferential reporting to more critical styles of journalism (Bauer and Gregory 2007). Chemical and biological warfare appears to be a part of, rather than exception to, this trend.

Within this context, it is fair to say that protests and parliamentary questions provoked the Ministry of Defence and Porton into press and TV appearances, 'helping' journalists with their stories, 'building a climate of informed opinion', and eventually holding Open Days. It is also worth noting that close monitoring and transcription of TV programmes was made possible largely because video recorders, while still very expensive, were becoming more commercially available by the late 1960s (Greenberg 2008).

In late July 1968 John Morris, as mentioned, was able to report back to Healey that he thought 'there are good signs that our more helpful attitude is having a good effect on the attitude of rational commentators.[88] Behind the scenes, this calm outward response can be contrasted with a sense of moral outrage at what was felt to be 'tendentious and unbalanced' coverage, a need to win the hearts and minds of the 'rational' protestors, finding methods to ignore 'irrational' protestors, personal attacks on scientists and their families and so forth – prompting high level intervention by Healey and others.

This process of information management erodes the boundaries between the world of secrets and the open world. I opened this chapter by suggesting that the metaphor of a sphere of secrecy does not do justice to the operation of secrecy in practice. Throughout this chapter, we have seen that the very existence of the Porton establishments provided a concrete geographical focus for discussions about secrecy. Indeed, many protestors were keen to stress that if Porton's work was truly defensive then there was no need for secrecy. In this respect, secrecy was more the direct target of protest than biological and chemical defence research. Likewise, the spokesmen for Porton Down were keen to try and show that secrecy was justified and also limited, with stress for example on the open publication of much of the work at the establishments. Press releases, television and other media appearances, the Open Days and so on were all gestures of transparency aimed at creating a favourable impression. Behind all of this, some things remained hidden. In Steven Roses' words in his *Twenty-Four Hours* interview 'what worries me is not the open day but what goes on the other 364 days'. In Empson's terms, introduced in chapter one, secrecy and openness are not so much opposed as fore-grounded and back-grounded at different times (Empson 2007). Or, drawing on a geographical analogy, we can say that openness exists as a zone between two others: the initial secret, what is then revealed about that secret, and what is held back and remains out of sight.

---

88 TNA, DEFE 11/998. JM to Secretary of State. Porton. (22 July 1968).

## Chapter 7

# Secret Spaces of Science: A Secret Formula, a Rogue Patent and Public Knowledge about Nerve Gas

Which is more dangerous: the four-page patent specification for VX nerve gas or Nesquik milkshake powder? In case the question seems frivolous, let me proceed with two vignettes, the first concerning Nesquik:

In the wake of the September 11 2001 bombing of the World Trade Center, and the anthrax attacks almost immediately afterwards, there were some 3000 hoax anthrax attacks across the USA. One hoaxer was 29 year old Terry Olson from Utah. Olson had drifted between jobs and then became unable to work after an accident in 1995. He suffered depression and spent his days watching television. Shortly after September 11, he took an envelope of junk mail, emptied the mail and added some sugar mixed with chocolate Nesquik. Olson then phoned the authorities and reported that someone had tried to send him anthrax in the mail. Shortly afterwards, Olson was arrested for the hoax. When British journalist Jon Ronson tracked him down, Olson had spent seven months in jail awaiting trial – and, if found guilty of threatening to use a biological weapon, he faced a maximum sentence of life without the possibility of parole.

> "What did they charge you with?" [Ronson] asked him.
> "Weapons of mass destruction," he said. "Life imprisonment."
> "You must have said to them that Nesquik and sugar aren't weapons of mass destruction," [Ronson] said.
> "I didn't say anything," he replied.[1]

He did not need to say anything. In the shadow of the World Trade Center and anthrax attacks, the authorities had construed innocuous Nesquik as dangerous. Curiously enough, when we turn to the case study that is the focus of this current chapter, the reverse happens with nerve gas:

---

1 Ronson, J., 'Hoax', *The Guardian Weekend* (5 October 2002: 20–24). Although the threat to use biological weapons carries a maximum life sentence in the USA, Olson was eventually charged with, and pleaded guilty to, making false statements to the FBI; this crime carried a maximum penalty of five years imprisonment. See Associated Press, 'Man Pleads Guilty to Anthrax Hoax', *The Daily Utah Chronicle* (14 November 2002).

On 5 January 1975 the British newspaper, the *Sunday Times*, published an article entitled 'Terror Risk as Deadly Nerve Gas Secrets Are Revealed'.[2] The piece reported that the patent on one of the deadliest known chemical warfare agents, VX, was now available in a number of public libraries. How, asked the reporters, had this secret been so irresponsibly breached? Anyone, including terrorists, could access and make use of the information contained in the patent. Within ten days, copies of the patent had been withdrawn but not reclassified, a government review of declassification procedures was announced, and in Parliament the Minister for Defence announced that 'Her Majesty's Government have never patented VX'. The implication was that nothing, or nothing worth worrying about, had happened.

Behind the closed doors of Government, a series of discussions between civil servants, scientists and politicians had taken place. Was this a dangerous patent? Was this knowledge really a secret? And ultimately, was this even a patent for VX nerve gas? A remarkable set of modifications of position occurred before the Minister arrived at his announcement. This chapter recounts the story behind the VX disclosures. In doing so, it addresses three questions pertinent to social studies of science that may appear at first glance unrelated, but are intimately connected: What makes knowledge dangerous? How does secrecy operate to help produce knowledge that is dangerous or otherwise? What happens when 'nothing happens'? Throughout, I want to develop the argument from previous chapters that secrecy cannot simply be regarded as a negative phenomenon that obscures knowledge, but is, instead, an active tool that allows governments to define reality through the exercise of spatial-epistemic power.

## *Dangerous Knowledge*

How such properties as danger, abusability and lethality are ascribed to particular military-related technologies has received scant attention within the Science and Technology Studies literature. Are these properties inherent in weapons, or are they only dangerous once associated with heterogeneous networks of other artefacts, expertises, social groups and goals? In an attempt to come to grips with this problem, Grint and Woolgar (1997) have posed the question 'what's social about being shot?' They argue that ascribing properties to weapons is firstly a social process, and as a corollary note how contextual interpretation is essential as a part of this process. Hence, the authors note, in the classic argument over whether guns kill people or people do, a typical sociology of technology position is to draw attention to certain guns in conjunction with certain people, rendering both dangerous. This is the main point of my opening vignette: not that the authorities somehow failed to recognise milk-shake powder; rather, that danger became a relational judgment attributed to Nesquik in conjunction with a hoaxer in the wake of anthrax bioterrorism.

2 Anon (Insight Team), 'Terror Risk as Deadly Nerve Gas Secrets Are Revealed', *Sunday Times* (5 January 1975: 1–2).

Picking up this theme, Rappert argues that while it seems intuitive to base a judgement of dangerousness on the effects of weapons, in practice weapon effects are contestable and so any assessment 'begs questions about what options are being compared, by what criteria, in relation to what circumstances and by whom' (Rappert 2001: 566).[3] And when 'abusability' is invoked in practice by actors attempting to condone or condemn technology, they are similarly combining interpretations and judgements about possible effects, the likelihood of users misusing a technology and its potential to serve particular ends (Rappert 2005, 2007b). This said, the general point, that degrees of danger are socially mediated ascriptions, is *not* the same as saying that all things are equally dangerous. It does, however, suggest that secrecy may have a demonstrable effect on these judgements.

In turn, these specific anti-essentialist arguments about the properties of weapons draw on more familiar arguments within Science and Technology Studies. Briefly re-capping these arguments moves us away from analyses of weapons *per se* and towards the patent for a potential weapon, the topic of this chapter. The question of whether or not a patent contains sufficient information to be dangerous immediately provokes a response that codified scientific knowledge is insufficient to replicate any experiment. Tacit skills and knowledge (Collins 1985, MacKenzie and Spinardi 1995, Polanyi 1958, Vogel 2006), a wider network or 'trans-epistemic arena' consisting of such people and things as funding agencies, audiences, regulatory bodies, reagent suppliers and so forth (Knorr-Cetina 1982, Latour 1983), and in more mundane terms a minimal level of resources and infrastructure are all components of 'doing science' in addition to any codified instructions. Additionally, patents do not follow the same writing conventions as journal articles (Myers 1995), although both generally narrate a thoroughly internalist version of knowledge production, in other words one that makes no reference to the role of context in its account of how knowledge comes about (Bowker 1994). The version of events, wherein patent claims are solely the result of theory and experiment within science, what Bowker terms a 'legal fiction', sits uneasily with contextualist versions of events where 'everyone really knows' that the personal, institutional, political, social, economic, historical and their ilk are equally important for establishing the credibility of knowledge claims. Both can be invoked by actors and analysts. Furthermore, as will be seen to happen with the VX patent, by setting the 'legal fiction' up as the one real version, other versions of events can be invoked to undermine the credibility of initial claims made about the secret patent.

### *How Does 'Nothing Happen'?*

In this case study, we will see how the media claim that a significant event – the breach of a dangerous secret – had occurred was steadily countered by Government

3 For detailed case-studies that demonstrate how assessments of weapon effects (or predicted effects) depend on social, political and organisational factors see Collins and Pinch (1998), Eden (2004), MacKenzie (1990), Vogel (2008a).

claims that nothing of the sort had happened. Yet, like dangerousness and secrecy, an 'event' – its boundaries and significance – are socially, politically and spatially mediated ascriptions. This point is apparent when considering, for instance, how the media singles out and creates news events to cover from a far wider potential selection of on-going scientific research (Gregory and Miller 1998). Conversely, 'non-events' can also be created. In order to follow and appreciate the significance such non-events where there is 'nothing happening', I draw on Lynch and Bogen's meticulous analysis of the Iran-Contra hearings (Lynch and Bogen 1996).[4] Lynch and Bogen show how the public hearings operated to produce and endorse particular versions of events and submerge alternative interpretations. In their analysis of this 'social production of history', and as an official version of events was assembled through the hearings, the authors also draw attention to the production of what might be termed 'anti-historiography' (in parallel with Galison's (2004) terminology of anti-epistemology introduced in chapter 1). In this respect, they note an observation of Luhmann's concerning the ability of the State to defuse controversy: 'From the outside, one can get the impression that the state bureaucracy is constructed as a social network with the main aim of ensuring that nothing happens when something does happen' (Luhmann 1994: 35).

Lynch and Bogen build on this point, commenting that from the inside, various processes contribute to the social production of 'nothing happening'. In their words, 'such nullity was a contingent achievement that encompassed an entire series of activities ... for concocting, shredding, and forgetting history' (Lynch and Bogen 1996: 260). Their account provides numerous examples of how documents (as well as speech and memory) were used to create an official version of, apparently unexceptional, events from a disparate and often contradictory range of sources and materials. In short, they argue, 'nothing' does not just happen, it is *made* to happen. So, with the VX patent under scrutiny in what follows, the transition from the newspaper claim that the patent had been released to the announcement in Parliament that VX had never been patented involved, as we shall see, a tremendous amount of such nullifying activity.

### *VX*

The nerve agent VX, one of a series of V agents, ranks as one of the most toxic chemical warfare agents developed during the 20th century. A drop of VX with the weight of a grain of rice, around 10mg, is sufficient to kill a person within minutes of being applied to the skin. These agents act by inhibiting acetylcholinesterase, an enzyme that breaks down the neurotransmitter acetylcholine. The subsequent build-up acetylcholine throughout the central and peripheral nervous system is

4 The scandal emerged in the 1980s when it came to light that there had been arms sales to Iran, despite a US Government embargo, and that the profit from the sales had been used to fund Contra rebels opposed to the Nicaraguan government (Lynch and Bogen 1996).

extremely dangerous as the victim's nerves continually 'fire', rapidly leading to paralysis and then death.

V agents originated from the close links established before and during the Cold War between the chemical industry and the UK Government's Chemical Defence Experimental Establishment (CDEE), at Porton Down in Wiltshire. In their search for new chemical warfare agents in the 1950s, CDEE scientists approached the chemical industry and requested information about any chemicals that were found to be too toxic for civilian use (McLeish 1997, McLeish and Balmer 2012, Perry-Robinson 1971). One of the first discoveries passed to Porton under this scheme related to research undertaken at Plant Protection Limited, a subsidiary of the British firms Imperial Chemical Industries (ICI) and Cooper, McDougall and Robertson (CMR). In his research on novel insecticides, Dr Ranajit Ghosh of Plant Protection Limited produced a toxic compound potentially useable as an anti-personnel weapon. Although attempts to turn this poison, eventually named Amiton, into a marketable insecticide floundered, largely as a result of its extreme toxicity, this self-same effect made it attractive to researchers at CDEE.

Amiton formed the basis for a new series of agents dubbed the V-agents, with the 'V' standing for 'venomous'. In 1955, Ghosh applied for a patent on one such nerve gas, the agent eventually named VE. Porton, again, was informed of the new chemical and of even more toxic agents in the V series. This information did not remain in Britain. Under their long-standing tripartite arrangement, the USA, UK and Canada routinely exchanged information on chemical and biological warfare research (Carter and Pearson 1996). Scientists at the CDEE passed information on the novel V agents to their counterparts at Edgewood Arsenal in the USA (McLeish 1997). It was the US scientists who made the further structural changes that resulted in the viscous, oily liquid agent coded VX (Croddy et al. 2001, Tucker 2007). The percutaneous toxicity of the agent, and its persistence in the environment made it an attractive military option; by the end of the 1960s, VX had been stockpiled as a key part of the US chemical warfare capability. Just a few years later in the UK, shortly after Harold Wilson's Labour Government had been elected with a slim majority of three seats, VX was to become the focus of media attention.

### *The Open Story: Terror Risk as Deadly Nerve Gas Secrets are Revealed*

On 5 January 1975, the *Sunday Times* published an article by its Insight team of journalists entitled 'Terror Risk as Deadly Nerve Gas Secrets Are Revealed'.[5] The article reported that the previously secret patent for VX had 'mysteriously' been removed from the secret list and the patent specification deposited in a number of public libraries. Alarmingly, the article noted that 'now anyone can go into the Patent Office in London and, with a minimum of research in the library, find the

5 Anon (Insight Team), 'Terror Risk as Deadly Nerve Gas Secrets Are Revealed', *Sunday Times* (5 January 1975: 1–2).

specification number for VX and be shown minute details of how to make it … a student with a little ingenuity could manufacture the gas in a university laboratory'.[6]

The authors, presumably with a mind to Middle Eastern terrorism and, closer to home, the bombing campaign by the provisional Irish Republican Army (IRA) that had culminated in November 1974 with a devastating explosion in a pub in Birmingham, then warned of the dangers of terrorists obtaining the declassified information.[7] The article also expressed bewilderment at the possible reasons for the declassification, pointing out that publication normally functions to protect commercial interests or 'boost the academic prestige of the inventor'. The article gave no indication of how the Insight team had found out about the patent. Some mention of the origins of the V agents at ICI was then made, before the author focussed on the patent itself.

The patent application had been filed on 21 June 1962 but, under the 1949 Patent Act which made provision for claims on inventions deemed 'prejudicial to the defence of the realm', had been kept secret by the then Secretary of State for Defence, Harold Watkinson. It had been declassified on 13 February 1974. Of course, as the journalists pointed out, the patent did not advertise itself as a claim on a nerve gas. Instead:

> The document submitted was well-disguised. The six foolscap pages, simply bearing the royal coat of arms, described in minute detail the entire process and gave the relevant chemical equations. But there was no hint that the process led to the deadly gas, VX. The document was blandly entitled 'Improvements in the manufacture of organic phosphorus compounds containing sulphur.' In a classic piece of military understatement the final products were described as 'usually highly toxic, as is now known. When mixed with suitable inert carrier material they may be used as insecticidal compositions'.[8]

While no mention of the patent number was made in the story, accompanying the article was a photograph of the front of the patent specification. The photographer had avoided taking a picture of the patent number, but included the application

---

6 Anon (Insight Team), 'Terror Risk as Deadly Nerve Gas Secrets Are Revealed', *Sunday Times* (5 January 1975: 1–2).

7 In a House of Lords debate on chemical weapons in June 1972, Lord Chalfont, a former Minister in the Foreign and Commonwealth Office, stressed the possibility that terrorists would adopt chemical weapons. *Hansard* (Lords) Vol. 331 c311–62 cited in *HSP CBW Events 1945–85* (HSP, SPRU, University of Sussex). The 1970s also witnessed the spread of small terrorist groups across Western Europe, most prominently at this time the Rote Armee Fraction or 'Baader-Meinhof gang'.

8 Anon (Insight Team), 'Terror Risk as Deadly Nerve Gas Secrets Are Revealed', *Sunday Times* (5 January 1975: 1–2).

number, from which anyone familiar with the UK patent system could locate the patent.

The next day the story was reported in the main daily papers.[9] Besides reiterating the main details of the *Sunday Times* article, these shorter pieces also added that Ministers from the Labour government would be facing a 'barrage of questions' in Parliament.[10] In particular, Bruce George, Labour MP for Walsall South in the Midlands, was tabling a Commons question on the released patent. The question, calling for a statement on the matter, would be put to the Home Secretary, Roy Jenkins, when Parliament reassembled. Once again the emphasis in these articles was on the patent for VX, the ease with which the patent could be accessed, explaining that this was a deadly nerve gas, and warning of the consequences if it was to fall into the hands of the IRA or other terrorists. Internationally, far shorter stories covering the release of the patent for VX appeared in the *Washington Post* and the *International Herald Tribune*.[11]

British journalists blurred the distinction between the formula and the process for making the nerve gas. The *Times* noted that '... the formula of one of the lethal nerve gases known as V-agents has been taken off the secret list. Any person can obtain a copy of the process by which it is made from the Patent Office'. In a similar vein, the *Guardian* pointed out in the opening paragraph of its article that 'the formula of the gas, named VX, has been public knowledge for two years'. Later in the article, this statement was heavily qualified by quoting the co-author of a study of chemical warfare, Julian Perry-Robinson from Sussex University. The six-volume study, issued by the Stockholm International Peace Research Institute (SIPRI), formed the most comprehensive public account of chemical and biological warfare issues, and according to Perry-Robinson, this work had promoted 'the declassification of the chemical structure of VX but not the method of production'.

At lunchtime, the BBC Radio 4 news programme, World at One, interviewed Labour MP, Bruce Douglas-Mann about the matter. Douglas-Mann claimed to have contacted the Minister of State for Defence, William Rodgers, and said that 'anyone contacting the patent office today would find they obviously cannot get the information'.[12] Indeed, as quietly as they had introduced the declassified specification, the Patent Office had taken the patent out of public circulation. As

---

9 Wright, P 'Questions on nerve gas to be put to ministers', *The Times* (6 January 1975); Anon. 'Patently Risky' *The Times* (6 January 1975); Anon. 'Killer Gas Formula off Secret List', *Daily Telegraph* (6 January 1975); Clark, D 'MP Demands Plug on Nerve Gas Leak', *Guardian* (6 January 1975); Anon. 'Nerve gas secrets "open to terrorists"', *The Sun* (6 January 1975).

10 Anon. 'Killer Gas Formula off Secret List', *Daily Telegraph* (6 January 1975).

11 Anon (Reuters) 'Fatal Gas Recipe Gets Out', *Washington Post* (6 January 1975); Anon (Reuters) 'Nerve-Gas Secret available in UK', *International Herald Tribune* (6 January 1975).

12 TNA, DEFE 13/823. Transcript of interview with Bruce Douglas-Mann MP, World at One.

one freelance writer recorded in his account of a hasty visit to obtain both the offending patent and a similar patent:

> I found the numbers for the two patents in less than one minute. But I also found that the relevant bound volume of printed patents had been removed from the library shelves and replaced by a loose-leaf box file. The two gas patents were missing from this file … I then ordered printed copies of the specifications but was subsequently informed that the patent had been 'withdrawn from circulation'. As a last resort I paid the requisite 15p, twice over, to inspect the Patent Office files on the two cases. These files which are normally public, ought to contain the original patent application typescripts … But they had also been withdrawn.[13]

At 19.00 hours on Monday 6 January, the Ministry of Defence (MoD) issued a press release that made some attempt to explain the appearance of the patent in the public arena:

> In fact it seems that a decision to de-classify information on VX was made in 1971 because Britain had no military requirement for it, and its properties were well known. Subsequently I understand that there has been widespread discussion of the formulae and manufacturing processes for nerve agents such as VX. This has been partly the result of a process of disclosure with a view towards the prohibition of chemical weapons. Against this background the de-classification of the patent application some months ago was seen as a recognition of realities making no significant new information available.[14]

The press release added that William Rodgers, Minister of State for Defence had, nevertheless, ordered a review of declassification procedures. Short articles followed in the *Daily Telegraph, Times, Guardian* and the *Sun,* all focussing on the decision to launch a review.[15] Only the *Daily Telegraph* and *Sun* added the detail that the decision regarding the VX patent had been taken 'because Britain had no military requirement for the gas and its properties were generally known'. A day later the *Times* added a brief article noting that a Pentagon spokesman had informed them that the US had declassified the formula for VX in December 1971 because it had 'by then become widely known'.[16]

---

13 Hope, A 'Patently Ridiculous', *New Scientist* (16 January 1975: 131–2).

14 TNA, DEFE 13/823. Ministry of Defence News Release. Given to Press Agencies at 1900 Hrs 6.1.75. VX Nerve Gas.

15 Anon. 'Nerve Gas Inquiry Ordered', *Daily Telegraph* (7 January 1975); Anon. 'Minister acts on nerve gas', *The Times* (7 January 1975); Anon. 'Chemical Weapons Review', *Guardian* (7 January 1975), Anon. 'Probe Over Nerve Gas Fears', *Sun* (7 January 1975).

16 Anon. 'Gas Formula Available in US', *Times* (8 January 1975).

By the end of the week, it was the turn of the science press to comment on the story. *Nature* provided a brief round-up of the story in the context of news from the USA.[17] Most of this article concentrated on the US Government finally ratifying the 1925 Geneva Protocol banning the first use of chemical and biological weapons. The *New Scientist* carried a more extensive commentary by Perry-Robinson who reiterated that the formula had been declassified in 1971. With respect to the patent reported in the *Sunday Times* and a similar one that had come to light in the interim, he added that 'only those who know that the symbol VX and a particular 50-word letter denote one and the same chemical will realise what the patents are about'.[18] This point highlighted a detail in the patent specification described by the *Sunday Times*. The patent did not mention VX, only a particular chemical formula and the process of synthesising the compound. Nerve gases were not mentioned either. The utility requirement of the patent was satisfied solely by a reference to the use of the chemicals as insecticides, which Perry-Robinson dismissed as 'a dangerous piece of nonsense'. However, the most dangerous disclosure in the patents, according to Perry-Robinson was the set of instructions for making VX:

> The two new patents give simple step by step recipes for making it in standard laboratory apparatus from commercial chemicals. No descriptions anything like as detailed have ever been published for VX. All that has been available is a set of chemical equations devoid of those essential practical details which only a chemist of exceptional experience and courage could provide for himself.[19]

Throughout the week, the *Sunday Times* had also received letters accusing them of irresponsible journalism. Only two of the letters were eventually printed, although a larger number survive in the archives of the Harvard-Sussex Program on Chemical Biological Weapons Armament and Arms Limitation. A patent agent and patent clerk both wrote in to say how easy it was to find the patent application from the information provided in the original article.[20] The patent agent, in particular, had no doubt that without the indiscretion of the *Sunday Times* an amateur would have been unable to identify and locate the relevant document. Another letter noted that 'what is dangerous is when the half-informed tell the half-witted how and where to

17 Norman, C 'USA Ratifies Chemical Warfare Protocol', *Nature* Vol.253 (10 January 1975: 82–3).

18 Perry-Robinson, J 'Behind the VX disclosure', *New Scientist* (9 January 1975): 50.

19 Perry-Robinson, J 'Behind the VX disclosure', *New Scientist* (9 January 1975): 50.

20 HSP [Harvard Sussex Program] Databank, Chronological Files, University of Sussex. Letter Drummond, P to The Editor (1975) *Sunday Times* (6 January 1975); Letter Webb, M to The Editor *Sunday Times* (7 January 1975).

find such information on the front pages of popular papers'.[21] Although, the writer added, some chemists could certainly build the necessary apparatus to make VX, 'considering the calibre of the modern student, and the woolly collection of theory now taught as chemistry, it is possible that some could not'. As mentioned, the newspaper published just two of the letters and their response was unapologetic: 'The *Sunday Times* omitted any mention of the specific patent number. Publications of our report led to a MoD inquiry and, we understand, withdrawal of the VX patent from easy availability'.[22]

And, as if to force the lid shut on the story, the paper also ran a brief article headlined 'VX gas is secret again'.[23] This reported that the MoD had removed the 'details of how to make VX' from the Patent Office within 24 hours of the *Sunday Times* reporting on their existence. With the recess over the following day, William Rodgers was left to face the barrage of Parliamentary questions promised in the *Daily Telegraph*. His responses received very little press comment. Three brief paragraphs by the *Daily Telegraph* reported that national security had been considered during the decision to declassify the patent.

As far as the press and academic commentary had been concerned in the previous week, this had been unequivocally a story about the public availability of a secret patent for VX nerve gas. Yet, in Parliament, when the Minister for Defence was asked what classification had been given to patent application No. 24022/62 and why it had changed, he responded with a different story. According to Rodgers, 'Patent Application No. 24022/62 does not relate to VX nerve gas but to improvements in the process of manufacture of a class of compounds of which VX is a member'.[24] And, ostensibly to clarify matters further, Rodgers responded to a further question the next day on the VX patent: 'Her Majesty's Government have never patented VX. The patent application made in 1962 covered certain improvements in the chemical processes for making the group of compounds which includes V agents'.[25]

Behind the scenes the claims made in the press had slowly, and secretly, been unravelling, although it was to take until May to fully resolve the situation. Within Whitehall, by the close of the episode, not only was the patent for VX no longer a patent for VX; the secret had never been a secret in the first place; and the formula and processes for making it were no longer prejudicial to the defence of the realm.

---

21 HSP Databank, Chronological Files, University of Sussex. Letter Urben, P to The Editor, *Sunday Times* (1975 no specific date).

22 HSP Databank, Chronological Files, University of Sussex. .Letter Drummond, P to The Editor (1975) *Sunday Times* (6 January 1975); and response: 'How to make nerve gas' *Sunday Times* (12 January 1975: 11).

23 Anon. 'VX gas is secret again', *Sunday Times* (12 January 1975).

24 *Hansard* 13 January 1975.Vol.884 col.20.

25 *Hansard* 14 January 1975. Vol.884 col.61.

*The Hidden Story*

The day after the *Sunday Times* announced the discovery of the rogue patent, the Private Secretary to the Minister of State for Defence, David Young, telephoned the Ministry's Director of Chemical and Biological Research, Ken Norris, with instructions. The Minster of State, William Rodgers, had decided to move rapidly and issue a statement. In a confidential note confirming the details of the phone conversation, Young asked for a history of the patent, including the reasons behind the original application and for the declassification.[26] The Private Secretary also stressed that he needed to know whether the disclosure had made any more information available to terrorists than had previously been in the public domain. In particular, he thought it was important to know whether the patent contained enough information for amateur chemists, again including terrorists, to obtain or manufacture both VX and means to deliver it to a target.

Young pressed for a response by 15.00 hours. In the interim, the MoD Press Office had been told to adopt a holding position in relation to the press. All that would be said was that 'an urgent investigation is in progress and a statement will be made later'.[27] If a persistent journalist were to try the Home Office, they would be pushed back to the MoD. Rodgers' superior, the Secretary of State for Defence, Roy Mason, had by this time also received a short memorandum from Bruce George asking for an investigation into the declassification and for a public statement on the matter.[28]

By the end of the day, Norris had put together a three page long, restricted circulation, set of background notes on VX. These notes were to form one of two written submissions from science advisors in the ensuing debate amongst Ministers and civil servants – therefore the contents are important for setting out the 'facts' as provided by state-sanctioned scientific authority. Norris first described what had been revealed previously about VX. Starting in 1968, during the Vietnam War, there had been an upsurge of public protest over chemical and biological warfare. By way of response, the Government had permitted open days at Porton Down and been represented at a 'teach-in' at Edinburgh. Norris noted that the 'general formulae for VX' had been disclosed at the Edinburgh meeting.[29]

During this time, researchers at SIPRI had embarked on their ambitious study of chemical and biological warfare. SIPRI had requested throughout 1970 for the VX formula to be declassified. And, Norris added, 'SIPRI had by this time derived

26 TNA, DEFE 13/823. Minister of State for Defence (PS/Minister of State D.E. Young) (6 January 1975).

27 TNA, DEFE 13/823. Minister of State for Defence (PS/Minister of State D.E. Young) (6 January 1975).

28 TNA, DEFE 13/823. Bruce George MP to the Rt Hon Roy Mason, Ministry of Defence (6 January 1975).

29 TNA, DEFE 13/823. Some Notes on the Security Classification, Production and Patents on VX. KP Norris. DRCB. (6 January 1975).

the correct formulae'.[30] The following year, the USA also requested the release of the VX formula 'because they wished to remove large quantities of VX stored on Okinawa and they were under an obligation to provide Japanese authorities with factual information on VX'.[31] In May 1971, the British Government complied with the request from the USA and the VX formula was declassified, although obviously not publicised.

Moreover, by this time, not just the formula, but also the characteristics of the agent were no longer classified as secret. As Norris put it:

> Further discussions within MOD came to the conclusion that VX should be regarded as a 'standard agent' and as the UK had no offensive capability it was agreed that the V agents could now … be regarded as coming into the category of agents whose properties were generally known. This meant that the chemical, physical and general physiological properties of VX became unclassified. At the same time the UK retained a classification on the effect of VX on military effectiveness.[32]

Norris continued with his chronology, summarised in Table 7.1, adding several other instances where the formula for VX had been disclosed. Then he explained the rationale for the British declassification:

> In October 1973 the file containing the Patent Application No 24022/62, i.e. referring to the manufacturing process for VX, came up for routine review. In the light of the previous discussions in MOD on the declassification of VX it was decided that there was no longer any justification for maintaining a SECRET grading. The normal process for patent publication and acceptance was then instituted and after publication on 13 Feb 1974 Letters Patent were received for Patent No 1346409 on 13 June 1974.[33]

The main thrust of Norris' account was to demonstrate that the secret was not so secret after all. After recounting the full chronology, Norris admitted to being bemused by the affair, claiming that 'in the light of these facts it will be clear that extensive information on the nature, properties and method of preparation of both VX and GB [another agent] have been available for a number of years

30 TNA, DEFE 13/823. Some Notes on the Security Classification, Production and Patents on VX. KP Norris. DRCB. (6 January 1975).

31 TNA, DEFE 13/823. Some Notes on the Security Classification, Production and Patents on VX. KP Norris. DRCB. (6 January 1975).

32 TNA, DEFE 13/823. Some Notes on the Security Classification, Production and Patents on VX. KP Norris. DRCB. (6 January 1975).

33 TNA, DEFE 13/823. Some Notes on the Security Classification, Production and Patents on VX. KP Norris. DRCB. (6 January 1975).

**Table 7.1 Summary of Open Information About VX, according to K. Norris**

| Date | Event |
|---|---|
| 1968–9 | Formula for VX given at Edinburgh 'teach in'. |
| 1970 | SIPRI 'derived' correct formula for VX. |
| 1971 | US request that UK declassify formula for VX as part of their operation to remove VX in bulk from Okinawa. |
| May 1971 | UK agree to declassify formula for VX. Further discussion leads to declassification of general properties of VX. |
| 1972 | Conference of the Committee of Disarmament study chemical warfare in detail. Includes discussion of 'methods of manufacturing nerve agents with a view to attempting to control the supply of precursors and of monitoring a nation's CW capability'. |
| 1972–3 | SIPRI begin publishing the six-volume 'The Problem of Chemical and Biological Warfare'. Includes 'extensive' information on nerve agents and a 'comprehensive bibliography'. |
| October 1973 | Routine review of patent application 'referring to the manufacturing process for VX'. Patent publication process initiated in Feb 1974. |
| November 1974 | Mid-West Research Institute prepares an account of the working of the US Newport Chemical Plant, giving 'considerable detail of the VX manufacturing process'. |
| April 1974 | Former SIPRI researcher, Julian Perry-Robinson presents papers on manufacture of G and V agents to Symposium on Chemical Weapons and Policy and Pugwash Chemical Warfare Workshop. |

from published sources. The current interest in the VX patent is therefore a little puzzling'.[34]

In conclusion, Norris addressed the specific questions that he had been requested to answer. The MoD had applied for the patent, he stated, because 'it is the general principle to protect the crown against possible exploitation of the invention. There were potential applications to the production of insecticides'.[35] The patent applications, while remaining secret for being prejudicial to the defence of the realm, had been routinely reviewed thus leading to the declassification decision. Finally, Norris added that competent chemists would be able to use the patent as a recipe for nerve agents but 'would be advised not to attempt to do so because of

34 TNA, DEFE 13/823. Some Notes on the Security Classification, Production and Patents on VX. KP Norris. DRCB. (6 January 1975).

35 TNA, DEFE 13/823. Some Notes on the Security Classification, Production and Patents on VX. KP Norris. DRCB. (6 January 1975).

the danger involved'.[36] Within hours, Norris' briefing had been condensed into two letters, to Douglas-Mann and George, as well as the press release quoted earlier.

## Making the Secret, that had never been a Secret, Secret Again …

As the following morning's papers announced that there would be a review of future declassification procedures, behind the scenes in Whitehall the Department of Trade, only recently responsible for the Patent Office, continued to express concern over the current declassification. The Patent Office had been placed in a politically embarrassing position and needed to manoeuvre out of it. One route would be to reclassify the patent as secret once more, although this decision could only be made by the Secretary of State for Defence on the grounds of the information being prejudicial to the defence of the realm. Accordingly, Anthony Hutton, Private Secretary to the Secretary of State for Trade, wrote to Young at the MoD:

> My Secretary of State fully understands that the information is now generally available and that restricting access to the patent itself would only marginally contribute to greater security. Nevertheless he has asked me to say that he regards it as quite intolerable and indefensible that such information should be freely available from HMG [Her Majesty's Government] and he would ask most strongly therefore that your Department should restore the classification as urgently as possible.[37]

Young saw the situation differently. Mindful that the patent had been released to 'at least a dozen' libraries in the UK, he noted bluntly in an internal memo that 'it is not practical politics to withdraw it'.[38] Patent applications had also been filed overseas. His response to the Department of Trade was somewhat more considered, equally negative and placed the onus for following through any recall of the patent firmly outside of the MoD:

> In the course of yesterday we considered the question whether we could take immediate steps to withdraw the patent on VX from public access … It seemed to us in these circumstances [the patent now widely available] that to attempt to withdraw the papers was not only a doubtful course of action but one not likely to succeed … However if you were to decide on this course … [i]n view of your Secretary of State's responsibility for

36 TNA, DEFE 13/823. Some Notes on the Security Classification, Production and Patents on VX. KP Norris. DRCB. (6 January 1975).

37 TNA, DEFE 13/823. Anthony C. Hutton (Private Secretary). Department of Trade (From the Secretary of State) to David E. Young (Private Secretary to the Minister of State for Defence) (7 January 1975).

38 TNA, DEFE 13/823. DE Young to DUS (Admin) PE. (7 January 1975).

> the Patent Office, Mr Rodgers assumes that you will have set in hand parallel, an urgent investigation of your own into how a decision to reclassify would be carried out and how effective it would be, bearing in mind particularly the fact that the patent is available in a number of overseas countries.[39]

Within the MoD, civil servants and politicians again turned to their scientific advisors for help on the looming problem of reclassification. This time, the information came from Ron Holmes, the Acting Director of Chemical and Biological Research. Holmes reported that a patent that 'described the general class of compounds to which the V-agents belong' was applied for by Ghosh in June 1955 and published in July 1958.[40] Later, in 1960, 'five methods for the preparation of compounds closely related to the V-agents were published with full preparative details' by Calderbank and Ghosh in the *Journal of the Chemical Society*. This paper provided information about Amiton, leading Holmes to then compare the chemical formulae of Amiton and VX. According to Holmes, 'the difference between the two is very slight in chemical terms and once the formula of VX were known the methods of Calderbank and Ghosh could be used with only slight changes of certain starting materials'. To drive home the point, Holmes declared that in fact 'V-type compounds' had independently been prepared, and their details published in the open literature, in seven different countries: Sweden (1957), West Germany (1958), USA (1956), Netherlands (1968), Yugoslavia (1968), UK (1972), and Czechoslovakia (1972).

Once again, the underlying message from the scientific advisors was that the wall around the secret had already been breached. Holmes pushed further. What exactly was the patent for? According to the media discussion, and debate in Whitehall up to this point, this was a patent for VX. Not according to Holmes, who wrote:

> the patent does not refer specifically to VX and as its title indicates is merely an alternative process ... Thus all of these patents deal only with generalised methods of manufacture of classes of compounds; they do not refer to specific agents and their publication does not represent a declassification of particular substances.[41]

Strictly speaking, Holmes' assertion was incontestable. Nowhere in the patent is VX mentioned. Nonetheless, the patent did claim specifically for a process for the manufacture of O-ethyl S-2-di-isopropylaminoethyl methyl-phosphonothioate

39 TNA, DEFE 13/823. DE Young to Anthony C Hutton (7 January 1975).

40 TNA, DEFE 13/823. Security Classification and Production of VX. R. Holmes. (8 January 1975).

41 TNA, DEFE 13/823. Security Classification and Production of VX. R. Holmes. (8 January 1975).

and for the product itself.[42] The claim was also made for the process for making a *general* class of compounds although, prior to the actual claim section in the specification and by way of example, the patent outlined how to make O-ethyl S-2-di-isopropylaminoethyl methyl-phosphonothioate. This short description provided codification of practical knowledge and conditions (concentrations, temperature, timing etc) needed to produce the compound.

By Friday 10 January 1975, the Holmes version of what the patent was for had started to circulate. In a letter copied to the Prime Minister, William Rodgers wrote directly to Peter Shore, the President of the Board of Trade (and Secretary of State for Trade):

> On the bigger issue the basic case for removing the classification from patent papers was that the general properties of VX had become widely known … It is also pertinent that the patent does not give the formula for VX, but covers the improved process of manufacturing agents in the V group, some of which are used in insecticides.[43]

Rodgers also took the opportunity to remind Shore that the number of patents in circulation had proliferated to some 20 libraries in the UK, around 70 abroad and now also to a number of industrial firms. Restricting the patents, he added, could be done in principle by libraries returning their patents and the Patent Office refusing to sell copies over the counter. Rodgers was quick to point out that this would have no legal backing and forthright about where such action might lead:

> I understand that the advice of the Treasury Solicitor is that the Controller [sic] of the Patents Office has no power under the Patents Acts to reapply the secrecy procedure to published patents. So any action would not have the force of law. The only way by which we could oblige all those people who have copies of the patents to return or destroy them would be to proceed against them under the Official Secrets Act. I am sure that you would agree with me that this is unthinkable … If we were to take steps of any kind to reduce the number of copies of the patent in circulation, we should be doing so in the full glare of other publicity. It would be widely known that our efforts could only be partly successful and we might look rather silly.[44]

Four days later, on 14 January, Rodgers drew on the Holmes version of events to declare to Parliament that Her Majesty's Government had never patented VX. At

42 The formula for VX remains declassified information by the US army (Department of the Army, 2005).

43 TNA, DEFE 13/823. Minister of State for Defence (William Rodgers) to the Rt Hon Peter Shore MP (10 January 1975).

44 TNA, DEFE 13/823. Minister of State for Defence (William Rodgers) to the Rt Hon Peter Shore MP (10 January 1975).

the same time Shore replied to Rodgers. Accepting that the information could never be completely recalled, Shore nonetheless wanted to restrict access to the patent. Officials in the Patent Office had suggested that they could 'do something quietly to this end' if the Secretary of State for Defence withdrew the release notices on the original patent plus the related patents that had since come to light. In the meantime, Shore informed Rodgers, the Patent Office and Science Reference Library were already restricting access to the patents.

By now, the Prime Minister, Harold Wilson, had requested further information on the patent. The resulting background information briefing from the Minister of State for Defence repeated much of the earlier story but with some additional details. The Prime Minister was informed of the 1958 patent on V agent production and of 'other open publications' since 1960.[45] On the patent specification that appeared in the *Sunday Times*, he was told that the War Office had filed the patent on 'an improved process for manufacturing agents of the V group' in 1962.[46] The note added that this was 'normal practice' by Government in order to protect the Crown from third parties independently taking out patents on government inventions. In the case of sensitive patents and at the behest of the originating department, the Prime Minister was informed, the patent application would be filed in the security section of the Patent Office without disclosure or publication of the information. Currently, there were around 90 UK patent applications held as secret in the Secretary of State for Defence's name.

The briefing note also provided additional details on the declassification process. VX had only been manufactured in the UK in small quantities; the US had manufactured it for use as a chemical weapon. The information on the progressive release of information about the V agents was summarised from the earlier information provided by Norris. When eventually, in 1971, the US requested the declassification of information on VX, it was agreed upon and simultaneously declassified in the UK, Canada, Australia, France and the Netherlands. Finally, 'under the annual review procedure the patent application concerning the V group of agents was considered for declassification in 1973. In the light of the declassification of VX and the increasing openness of information in the general areas and after consultation with the US authorities, it was decided that the information in the patent application could be declassified'. So, although VX is mentioned throughout this briefing, by this stage of the discussion the Minister of State for Defence was careful to avoid stating to the Prime Minister that the patent was for anything as specific as one particular nerve agent.

45 TNA, DEFE 13/823. Background Note. Minister of State for Defence to Prime Minister (14 January 1975).

46 The War Office was a predecessor to the MoD.

*A Meeting at the Patent Office*

In order to resolve the position on reclassifying the patent, the Department of Trade scheduled a meeting for Wednesday 15 January 1975. At the meeting, the assembled officials were informed that there were now seven published patents that needed to be considered. If steps were taken to withdraw the release notice on these patents, the printed copies would need to be withdrawn from sale, public inspection of the originals would be refused and UK libraries would have to be instructed to prevent further access to copies in their possession. On the other hand, 'it would be inexpedient to make a similar direction to others known to hold a copy e.g. the *Sunday Times*'.[47] And, if that was not problematic enough, a further 25 copies had been purchased by 'unidentified persons'.[48]

Two legal experts at the meeting outlined the legal position on reclassification, admitting that it would be 'straining somewhat the wording of the Patents Act'.[49] Only the Secretary of State for Defence had the power to 'lawfully declare that the publication of the information in question was now prejudicial to the defence of the realm'.[50] However, the consensus of the meeting was that withdrawal of the release notice post-publication would be unlikely to be backed by the courts. Representatives from the MoD added that there were practical reasons not to withdraw the patent. These included the wider availability of information on the nerve gases; the need to consult with the US and Canada before withdrawing the information; and the problem that, while reclassification might deter some, the 'withdrawal of the notice would not be effective vis a vis the evil-minded'.[51]

The officials also considered what would happen if the patent was not reclassified but instead its inspection and sale was restricted to 'identified people'?[52] This course of action was considered practicable, and possibly even beyond legal challenge. But it was also felt to be almost useless in protecting the information from a determined terrorist. Additionally, in notes written up the day following the

47 TNA, DEFE 13/823. Note of Meeting at the Patent Office on 15 January 1975. Subject: Nerve Gases. J.D. Fergusson (Patent Office) (16 January 1975). Attended by representatives of Department of Trade and Industry; Ministry of Defence; Treasury Solicitor's Department; Science Reference Library.

48 TNA, DEFE 13/823. Note of Meeting at the Patent Office on 15 January 1975. Subject: Nerve Gases. J.D. Fergusson (Patent Office) (16 January 1975).

49 TNA, DEFE 13/823. Note of Meeting at the Patent Office on 15 January 1975. Subject: Nerve Gases. J.D. Fergusson (Patent Office) (16 January 1975).

50 TNA, DEFE 13/823. K.T. Nash (DUS(Admin)PE) to PS/Minister of State. VX: Meeting of Officials at Department of Trade (17 January 1975).

51 TNA, DEFE 13/823. Note of Meeting at the Patent Office on 15 January 1975. Subject: Nerve Gases. J.D. Fergusson (Patent Office) (16 January 1975).

52 TNA, DEFE 13/823. Note of Meeting at the Patent Office on 15 January 1975. Subject: Nerve Gases. J.D. Fergusson (Patent Office) (16 January 1975).

meeting, the *New Scientist* article 'Patently Ridiculous' is mentioned as throwing the legality of such restrictions into doubt.[53]

The general feeling by the close of the meeting was to favour doing nothing. Accordingly, in the MoD representatives' record of the meeting, the consensus was that:

> In the end, the factors which remained uppermost in the minds of officials were that withdrawal of the release notice would not lead to any effective restriction on malefactors; that there would inevitably be further delay before it could be withdrawn; that up to seven patents could be involved, with therefore an increased chance of stirring up interest and trouble; and that the legal position of the Government, if the matter ever went to law, would have some noteworthy vulnerable spots. As regards enforcement of any restrictions there was a decided risk of embarrassment if the authorities refused HMG permission to prosecute, or if the prosecution was brought but failed or resulted in a nominal penalty; and of course any legal proceeding would inevitably result in publicity.[54]

On 17 January, the private secretary to the Minister of State for Defence wrote to his Minister that 'we are now in a position to take and announce final decisions'.[55] The patent should not be reclassified. Information had been freely available for almost a year; it would be 'impossible' to persuade the Attorney-General to prosecute anyone publishing the information and 'in any event such a course would … be likely to expose us to ridicule'. Moreover, copies would continue to remain in private hands. With action short of reclassification, this would not deter terrorists and 'we again run the risk of ridicule and would also have to argue that on the one hand the decision to declassify had been correct while on the other we were taking steps to minimise "the damage".'

Armed with this information, Rodgers wrote to Roy Mason, Secretary of State for Defence, in order to bring 'the discussion over VX to a conclusion'. Although the original decision to declassify the information on the V agents was justified in 'normal peacetime circumstances', Rodgers added that 'with hindsight, however, it was probably unwise'.[56] The reason being that 'even if the additional risk of making such information readily available to the terrorist had been regarded as minimal, it is important at a time like this not to undermine public confidence'. That said, Rodgers added, although it would be ideal to reverse the decision, this could not be

53 TNA, DEFE 13/823. K.T. Nash (DUS(Admin)PE) to PS/Minister of State. VX: Meeting of Officials at Department of Trade (17 January 1975).

54 TNA, DEFE 13/823. K.T. Nash (DUS(Admin)PE) to PS/Minister of State. VX: Meeting of Officials at Department of Trade (17 January 1975).

55 TNA, DEFE 13/823. PS to Minister of State for Defence (17 January 1975). VX.

56 TNA, DEFE 13/823. WTR (Minister of State for Defence) to the Secretary of State for Defence (20 January 1975).

done: 'in the first place, we should be closing the door after the horse had bolted; in the second, it could not be closed but would be left very obviously ajar'.[57]

Rodgers then proceeded to outline the reasoning put to him a few days earlier, concluding that 'we should do nothing, resting in any further public discussion both on the original case for taking the V agents off the secret list and on the present review of procedures'.[58] Rodgers suggested some sort of statement could be made in the form of a response to a well-placed parliamentary question. The Secretary of State soon responded to Rodgers with his agreement and explained:

> that our position is defensible on the grounds both that the formula for VX has been relatively easily accessible to the public for some time and that HMG's patent does not give the formula. He agrees that we might be criticised for having facilitated public access to a dangerous process. He considers it important, therefore, that the statement should aim at allaying doubts on this score, perhaps by emphasising the general effectiveness of our declassification procedures and the additional safeguards that have now been incorporated into the machinery.[59]

Two days after Rogers' memorandum, a confidential letter, containing the well-rehearsed arguments against declassification, was dispatched from Roy Mason to Peter Shore, Secretary of State for Trade.

### *Further Protest*

Shore did not respond directly to Roy Mason. Instead, he wrote straight to the Prime Minister. In the letter, Shore reminded the Prime Minister that he carried departmental responsibility for the Patent Office and that, while recognising the partial effectiveness of any action to declassify the nerve gas patents, some action should nevertheless be taken. The embarrassment to the Government of having this information on display was paramount as Shore wrote: 'It is publicly indefensible that such harmful information should be so easily accessible from Government sources, at a time when our society is under attack from extremists of many kinds'.[60] It is worth noting here a conflation, discussed in Chapter 1, of the usual distinction often made between legitimate secrecy to maintain national security and illegitimate secrecy to hide embarrassment (Gibbs 1995, Shils 1956).

---

57 TNA, DEFE 13/823. WTR (Minister of State for Defence) to the Secretary of State for Defence (20 January 1975).

58 TNA, DEFE 13/823. WTR (Minister of State for Defence) to the Secretary of State for Defence (20 January 1975).

59 TNA, DEFE 13/823. Letter AJ Cragg (APS/Secretary of State) to PS/Minister of State. VX. (22 January 1975).

60 TNA, DEFE 13/823. Letter Peter Shore to Harold (Wilson), The Prime Minister. Nerve Gases. (24 January 1975).

At the Prime Minister's behest, a special committee of Ministers was formed to deal with the nerve gas problem. The committee, titled MISC67, consisted of the Lord Chancellor in the chair, the Defence Secretary, the Trade Secretary, the Attorney General, and representing the libraries where the patents were held, the Minister for the Arts and Parliamentary Under Secretary of State (Health, Education and Social Work) of the Scottish Office. The first MISC67 meeting, on 6 February 1975, simply produced a request to the Attorney General for his expert opinion on the matter.[61] It took six days for him to prepare a statement outlining one possible course of action. The statement reasserted the legal impossibility of reclassifying the patent on the grounds of being prejudicial to the defence of the realm. This provision for maintaining secrecy only applied when the application for the patent was initially made, certainly not once it had been published and released into the public domain.[62] Although the patent rules allowed for inspection of the patents, the Attorney General suggested a further rule to restrict access to sensitive patents.[63] The draft version suggested that at three months or more past publication, further inspection of a patent could be refused if a 'competent authority … has certified that further inspection thereof could in his opinion be prejudicial to the safety of the public'.[64]

As civil servants were quick to point out, 'the formulation "prejudicial to the safety of the public" is of course very different from "prejudicial to the defence of the realm", and it seems that the "competent authority" for this purpose could be any one of several Secretaries of State'.[65] The Attorney General recognised that this now widened the powers of the Government to maintain secrecy. In his written opinion, he noted that the rule could be 'open to serious criticism' if a patent holder claimed for infringement yet the person claimed against was unable to consult the original patent.[66] Additionally, he wrote, 'only if it was thought that the secrecy would be of real value should the Rule be introduced. If introduced it should be subject to reasonable safeguards'.

---

61 MISC67 also considered re-formulating the patent so that it was limited to harmless applications. The advice of the civil service solicitors was that this would have been technically difficult and amending a published patent would have been legally impossible. See TNA, DEFE 13/823. Initialled KN. Position: DUS(ADMIN)(PE) to APS/S of S (17 February 1975).

62 TNA, DEFE 13/823. Patents Act 1949. Opinion of the Attorney General. (12 February 1975).

63 The patent rules are the secondary legislation that govern the procedures for implementing the Patents Acts.

64 TNA, DEFE 13/823. Patents Act 1949. Opinion of the Attorney General. (12 February 1975).

65 TNA, DEFE 13/823. Initialled KN. Position: DUS(ADMIN)(PE) to APS/S of S (17 February 1975).

66 TNA, DEFE 13/823. Patents Act 1949. Opinion of the Attorney General. (12 February 1975).

With this shift of emphasis Mason conceded, and his assistant private secretary wrote internally, that 'the Secretary of State … feels now the MISC67 committee has ruled in favour of taking action to restrict public access to the patent we should not press our opposition too strongly'.[67] Advice from civil servants had already stated that the Defence Secretary might 'now wish to leave the running to the S of S [Secretary of State] for Trade, who may still be expected to see real value, admittedly in political terms only, in the restrictions'.[68] Accordingly, a letter was sent from Mason to Shore stating that the issue of whether or not to introduce the new rule was a matter for him to decide in consultation with other colleagues. In addition, the prospect of the new rule meant that the MoD could safely discard the matter. On this point, Mason informed Shore:

> As I see it, the question of my re-classifying the patented information as being prejudicial to the Defence of the Realm does not now arise; and I take it that if a new rule was introduced, a number of Secretaries of State would in law be competent to certify that further inspection of the specification in question could in their opinion be prejudicial to the safety of the public.[69]

MISC67 met a second and final time on 18 April 1975. The committee members agreed to pursue the new rule for implementing the 1949 Patent Act and formulate it in such as way as to reduce the chance of any putative legal challenge. In early May the wording was agreed and Shore wrote to the Lord Chancellor that 'when the rule has been approved, I shall then put in hand the recall of copies of specifications from various libraries and have the appropriate machinery for monitoring inspection set up in the Patent Office'.[70]

The final form of the rule maintained, as in the draft, that the inspection could be refused for patent specifications deemed by the relevant Secretary of State to be 'prejudicial to the safety of the public'. A small proviso was added that the secret could be kept for a maximum of twelve months before being subject to review.

And, by way of a postscript, two years later a radical overhaul of the British patent system took place in order to bring it in line with European patent law (O'Dell 1994). The passage of the 1977 Patent Act through Parliament was incredibly stormy, with several hundred amendments tabled in both Houses. The draft Act maintained the Government's right to keep secret those patent applications judged prejudicial to the defence of the realm and additionally those deemed prejudicial to the safety of the public. When this secrecy clause was questioned in the House

---

67 TNA, DEFE 13/823. AJ Craggs APS/Secretary of State to DUS(Admin)(PE) (19 February 1975).

68 TNA, DEFE 13/823. Initialled KN. Position: DUS(ADMIN)(PE) to APS/S of S (17 February 1975).

69 TNA, DEFE 13/823. From Roy Mason to Peter Shore. VX. (19 February 1975).

70 TNA, DEFE 13/823. Peter Shore to Lord Elwyn-Jones, The Lord Chancellor. Nerve Gases. (8 May 1975).

of Lords, the government whip, Lord Oram, pleaded ignorance on its origins or reasons for inclusion in the bill. The clause was duly voted out of the bill by an overwhelming majority (98 to 44), and only reinstated at the last moment in the Commons. O'Dell (1994: 134) records that the 'Earl of Halsbury remarked upon reinstatement of the secrecy clauses in the bill "we can go on living with it for the time being".' In its own quiet way, the VX fiasco resulted in a shift in patent law and the subtle extension of the powers of the Government to impose secrecy.

## Secrets, Patents and Knowledge

### *Different Patents?*

During the episode various readings of the patent were articulated, by different groups in different places. Each version drew on different combinations of features to constitute an identity for VX. Key features that differed were: the relationship between 'essentialist' properties of VX, such as the formula, and the know-how deemed necessary to make the nerve agent; the degree of revelation that was deemed to have occurred as 'the secret' was differently constructed; and the presumed intent and ability of putative abusers.

Accordingly, different readings placed emphasis not so much on whether additional tacit knowledge was needed to make use of the patent, but rather on whether the patent added anything substantially new. For Perry-Robinson, writing in the *New Scientist*, the patent added to the previously 'devoid' formulas by codifying the tacit skills of an 'exceptional' chemist. Alternatively, the initial response within the MoD was that because the formula and properties had already been declassified, and reported outside of sanctioned military and government spaces, then whatever was in the patent added nothing new. Once under pressure from the Ministry of Trade to reclassify the patent, the MoD position shifted in accordance with advice from Holmes. The patent was read as a repository of knowledge about manufacturing processes, but no longer specifically for the properties (or even formula) of VX.

When sparring with the Ministry of Trade, the representatives from the MoD were keen to emphasise secrecy as pragmatic. Re-classification would not retract knowledge, but instead would serve to call into question the original declassification decision, thus reconstituting it as a noteworthy rather than routine event. In the Ministry of Trade, secrecy was construed as part of the identity of responsible government, hence the focus of officials on ensuring the legality of reclassifying the patent, coupled with their emphasis on the availability of even this minimal information to 'undermine public confidence' and when 'our society is under attack from extremists of many kinds'. Finally, insofar as the media coverage was homogeneous, it was marked by a conflation of formula, know-how and intent. So, in the initial Insight report, there was a smooth narrative movement from 'anyone' finding the formula in the public space of the Patent Office, to the knowledge and capabilities of an 'ingenious student' in a university laboratory with no specified motive, to a determined terrorist with an assumed reason for wanting to utilise specifically a chemical weapon.

So, these interpretations of the patent each presented different ideas of what the agent was, how its properties and manufacture were codified, where the practical skills in its production resided, how it intersected (and had historically intersected) with public domains, and how the public interest (and by implication the public) was itself constructed. With secrecy allowing the creation and persistence of different readings in different places, secrecy acted as a spatial-epistemic tool. As knowledge about VX and its 'context' was thus differently co-produced in multiple readings of the patent, each constituted 'the event' as something worth worrying about, or otherwise, and consequently suggested different possible ways of dealing with the release.

### *The Ministry of Defence*

The focus of this narrative has been the transition within the MoD from responding to media claims about VX, to the contradictory claim that this was not the patent for VX. Within the confines of secrecy, the status of the patent had been altered so that, insofar as the MoD, and later the Government, could claim that there was nothing to be concerned about, nothing appeared to have happened. In order to move between the VX patent as a dangerous security leak and the patent as relatively innocuous, three key claims were made and then rebutted. The claims were: that a secret had been inappropriately released (implying that the keepers of the secret were untrustworthy); that this was a patent for the nerve gas, VX; and that the secret was prejudicial to the defence of the nation. Some of the refutation of these claims took place publicly; in particular the statements made to the House of Commons by the Minster for Defence that HMG had 'never patented VX'. The majority of the unravelling of the public accusations took place behind closed doors – in meetings, phone conversations, memoranda and minutes.

### *Was this a Secret?*

> 'Extensive information on … VX … [has] been available for a number of years from published sources. The current interest in the VX patent is therefore a little puzzling'

The VX patent story appealed to various news values, the aspects of any story that render it newsworthy (Gregory and Miller 1998). These news values included a revelation, bad and risky events, and a focus on events happening contemporaneously. Not only was this story reported as a story about dangerous knowledge, but it was dangerous and truant knowledge that had escaped now, in the light of growing concerns over terrorist activities. Additionally, the news depended on this secret being construed as a bounded object, contained in space and time and previously only available within a restricted space. Behind the scenes at Whitehall, the opposite occurred as the secret was being transformed into a non-secret. Written documentation of 'the facts' was requested from state-sanctioned scientists twice in the course of the incident: Norris' background notes produced the day after the *Sunday Times* story appeared, and two days later, Holmes' note pointing out that the patent did not specifically refer to VX. Both were produced days before the announcement in Parliament that Her Majesty's Government had not patented VX.

The status of the patent as a singular secret was challenged in both memoranda by dispersing the knowledge spatially and temporally. Norris' memo outlined prior instances where the formula for VX had either been guessed or revealed, and noted that this, as well as revelations about the characteristics of the agent, had led to a separate process of declassification. In this sense, the secret had been bureaucratically downgraded and no longer labelled a secret even if this matter was not publicised. The precedent set by official declassification of information pertaining to VX was invoked as the reason why the patent was finally granted. Holmes, likewise, provided a history and geography for the supposed secret, noting that V-type compounds had been prepared in seven different countries between 1957 and 1972. So, for both scientists, there were certainly boundaries to the secret but they were far more temporally and spatially dispersed than had hitherto been portrayed in the current crisis.

In each case the secret as a bounded entity was juxtaposed against the non-secret as geographically and historically dispersed. Secrets, I have argued throughout this book, generally encompass social arrangements that are more complex. Secret cultures, indeed, are often spatially isolated but that secrecy is not fixed to particular 'technologies of secrecy'. For the MoD, a secret was co-terminus with being confined within the apparatus and arrangements designed for keeping secrets. So, when trying to demonstrate that the secret was never really secret, compiling a chronological list of the snippets of information that had emerged in dispersed locations about the V-agents – regardless of who would have had access to obtain and make practical sense of that information – was taken to mean that the secret had already been released in its entirety. Space and knowledge were, in this way, intimately intertwined.

### *What was the Patent for? 'Her Majesty's Government have never Patented VX'*

The claim that the patent was not the patent for VX appears to have originated, in written form at least, in Holmes' memo of 8 January 1975. Here, the Acting Director of Chemical and Biological Research claimed that because the patent did not mention VX specifically then it was not a patent for VX. Instead, he claimed, that the various patents dealt only with generalised methods of manufacture of classes of compounds.

We can only conjecture what was happening here. First, Holmes could have been (by his own standards of judgement) mistaken for any number of reasons: lack of knowledge, misreading of the patent, haste and so on. As a second possibility, Holmes was being disingenuous, performing a civil service sleight of hand. A third possibility is that Holmes genuinely believed that this was not a patent for VX – and no amount of counterfactual coaxing would have persuaded him otherwise. As just discussed, it would be mistaken to regard patents as having determinate meanings free from interpretation and, regardless of the reason for Holmes' claim, the notion that 'this was not a patent for VX' became the preferred (and unchallenged) reading of the patent within and beyond the MoD.

The consequence of Holmes' memorandum was that the 'two worlds' that Simmel (1906) associated with secrecy became manifest here – one where the patent was taken to be for VX, one where this was taken not to be the patent for VX. Secrecy meant that it was relatively easy for those in authority to produce their version of events as legitimate. Indeed, within the closed spaces first of the MoD and then Whitehall more generally, present knowledge of the danger (or lack thereof) solidified as the history of events (of the times and places where the V-agents were manufactured or known about) was recorded and reported. The anti-epistemology of secrecy discussed in Chapter 1 worked in this instance not so much as Galison claims, to cover up knowledge, but to actually re-constitute it. Short of a putative group of terrorists taking the offending patent, working through it and launching a successful attack with the agent, it is difficult to see how the secret Whitehall version of what the patent was for could have been countered. Indeed, as I now discuss, it appeared easier for Holmes to change the chemistry of the patent than for civil servants and Ministers to change the law and reclassify the patent.

## Is this Dangerous Knowledge?

In the course of the incident, during discussions over whether or not the patent could be rendered secret again, the Prime Minister had been briefed with the original reasons for keeping the patent application under cover. Patents were normally taken out on Government-funded inventions and, as in the present case, a sensitive patent would not be granted but the application held secret at the patent office to be produced should a third party independently attempt to apply for a similar patent. The power to hold the patent application under lock and key rested with the Secretary of State for Defence and could be exercised if the patent was deemed 'prejudicial to the defence of the realm'. From the Department of Trade perspective, restricting access to the patent 'would only marginally contribute to greater security', the MoD concurred that withdrawal would not deter the 'the evil-minded'. The concern of Peter Shore, speaking for the Department of Trade, was that the Government via Shore would be regarded as irresponsible for allowing this information to escape. Yet within the MoD, trying to recall the secret was deemed impractical, with no force of law behind it. It would be a half-hearted measure, yet change a routine declassification into a highly visible mistake.

Yet, as I have described, a trade-off remained between the political embarrassment of being associated with the leak, and the conviction that 'withdrawal of the notice would not be effective vis à vis the evil-minded'. This trade-off was achieved first by the MoD avoiding being forced to reclassify to secret what was now deemed, through Holmes and Norris' accounts, the 'not-so-secret'. By way of a compromise, the patent rules were instead widened to allow any Secretary of State, not just Defence, to restrict access to a patent on the grounds of being 'prejudicial to public safety'. The Ministry of Defence was perfectly happy with this compromise, with Shore noting that 'the question of my re-classifying the patented information as being prejudicial to the Defence of the Realm does not now arise'.

## Conclusion

This case study builds on my plea in Chapter 1 to take space and knowledge seriously when thinking about secrecy. Spatiality is not simply metaphorical in the production of knowledge; it can be shown to count in non-trivial and consequential ways. I have maintained that secrecy as a spatial-epistemic tool of governance ties together the three questions I posed from the outset. Knowledge is constructed as dangerous within particular spaces. The patent was construed as dangerous in the print media and beyond but eventually less so within Whitehall. How did secrecy operate? Secrecy created an opaque space where the Government version of what knowledge lay within, and what was known about, the patent retained legitimacy. And, by re-conceptualising what had been secret as public knowledge – or in Dennis' (1994) terms by re-configuring the boundaries in an archipelago of open and closed knowledge – this version served to underwrite claims that nothing had happened.

The discussion over the patent opened up into a whole raft of contingent questions: Is this patent dangerous? Is this patent for VX? Is the patent not VX but still dangerous? Who knows (and who knew) what was in the patent? Could a terrorist make VX? Would a terrorist make VX? Is Her Majesty's Government prepared to take the risk? Is Her Majesty's Government prepared to be identified as the source, should it happen – even if it is available elsewhere? As answers were given – or frequently assumed – for such questions through different readings of the patent, the initial set of claims headlined as 'Terror Risk as Deadly Nerve Gas Secrets Are Revealed' was defused. Within Whitehall, 'terror risk' was minimised, a determined terrorist would get the information anyway; the knowledge was no longer 'prejudicial to defence of the realm' but 'prejudicial to public safety'; 'deadly nerve gas' was no longer VX but the general class of V agents; and 'secrets' were no longer secrets but had already been revealed.

Answers to the many questions raised by the patent were therefore not simply about its identity, and certainly not just about the legitimacy or otherwise of a bureaucratic set of classification procedures. At stake was the authority of the MoD, and its scientific advisers, as arbiters of what counted as sensitive and classified information, and therefore what was to count as dangerous knowledge. At the same time, secrecy was used as a spatial-epistemic tool – with location, within different parts of Whitehall and beyond, as key in constituting and sustaining different readings of the patent.

In order to frame answers to the patent questions, and thus ensure that 'nothing happened', judgements and assumptions had to be made in secret and about secrecy. I argued above that the secret was treated at various times by both journalists and the authorities as a bounded entity, restricted within the pages of the patent document. This, in part, involved an adjunct conflation of the formula and the tacit knowledge (and other resources) needed to make VX. Additionally, by Holmes and Norris demonstrating that the secret patent did not conform to the 'legal fiction' of knowledge contained within the bounds of the four-page specification, this same demonstration was taken to imply that there had never been a secret.

Dealing with secrets at a normative level also involved civil servants, Ministers and scientists having to judge what others already knew and to attribute expertise and motives to others. Just as in the *Carella* incident discussed in chapter 3, where the public were presumed to possess considerable abilities to penetrate secrets based on small revelations, whether or not the nerve agent patents were deemed risky thus depended on various actors constructing the identity and capacities of the enemy, of those who would breach the secret. The 'determined terrorist' was regarded as being able to put together and understand the significance of the snippets of information that had preceded the release of the patent. The 'evil minded' were construed as being undeterred by any moves to reclassify the patent. Explicitly, Holmes noted that prior publication of information on the V-agents meant that once the specific formula for VX was known it would be easy to use this prior information to make this nerve agent. This, however, begs the questions of 'known by whom and how?'

As this case study demonstrates any answer to this question is equally an assessment about the potential uses of codified research outputs, such as patents, that involves someone having to decide what is dangerous in relation to the characteristics, motives, expertise and resources of putative wrongdoers. In this respect, a patent for a nerve gas only becomes dangerous when allied with particular expertise and resources. Some of these resources might be readily accessible, others quite scarce. Additionally, the patent would need to be allied with particular intentions, which in turn would have resource and expertise implications. Put bluntly, production on a large-scale intended for State use and military weaponisation requires insertion into a different network of skills and resources than 'kitchen-sink' production by anyone intent on primarily creating social and economic disruption.

Yet when the topography of knowledge is not a smooth and seamless landscape, and different knowledges operate in restricted and isolated spaces, then it becomes important to think about who is equipped and who gets to say what is or is not dangerous? At the very least, placing the burden of judgement onto singular, relatively isolated, communities left them to fill their gaps with assumptions that other communities may well have been better placed to challenge and revise.

# Chapter 8
# Opaque Science

In his analysis of blackmail as a prevalent theme in the nineteenth century novels of George Eliot, literary scholar Alexander Welsh argues that particular manifestations of secrecy, in his case secrets peculiarly susceptible to blackmail, are tied to historical circumstances (Welsh 1985). Against a backdrop of Victorian society, increasingly connected by new modes of transport and with wider access to the postal system, Welsh maintains that the nature of secrecy changed. Trust became paramount in an economy where more and more business deals operated at a distance and, for instance, someone filling a vacant job post might have to substitute trust for knowledge about a potential employee. At the same time, urbanisation in this period meant that while people lived in ever closer proximity, they might know less and less about each other. Finally, argues Welsh, with the rise of representative government, less dependent on custom or inheritance, the selection of who deserved to vote or to represent voters made a political career more sensitive to unsavoury revelation and blackmail. In short, industrialisation facilitated a particularly widespread and intense concern with secrets connected to blackmail.

The case studies in this book concerned the secrecy surrounding biological and chemical weapons in a peculiarly Cold War context. Undoubtedly this was a particular type of paranoia-tinged secrecy, linked to deadly knowledge obtained by a scientific elite, yet deeply connected with the power of the state. It is not difficult to place this manifestation of secrecy in a historical context of the steady folding together of scientific and military concerns throughout the twentieth century (Roland 2003), together with the post Second World War re-positioning of many scientists from their perceived role as intellectuals or servants of the state to potential security threats (Badash 2003, Goodman 2005, Thorpe 2002, Wang 1998). This said, and with all the usual caveats that history is not an instruction manual for the present, there may be certain enduring features of secrecy arising from the observations made throughout this book that we might be sensitised to when considering more recent events.[1] Let me illustrate this point with a few examples before turning to some more general remarks about secrecy.

Within a few years of the September 11 2001 terrorist attack on the twin towers of the New York World Trade Center, and the subsequent attacks using letters filled with anthrax (Cole 2003, Guillemin 2011), debate surfaced about

---

1 On the relationship between historical analysis and either contemporary social theory or policy analysis, see, Berridge (2003) Burke (2005), and Parliamentary Office of Science and Technology (POST) (2009).

the regulation of scientific research through secrecy. In relation to biological weapons this debate became focussed on the potentially malign use of scientific discoveries made within non-military settings (e.g. Rappert 2003, Royal Society 2005, Somerville and Atlas 2005). One contested proposal was that the USA should employ new security categories, such as 'sensitive but unclassified', as a way of regulating scientific information. In their opposition to this proposal, many scientists and scientific societies appealed to openness, transparency and freedom of information exchange as hallmarks of the scientific process (Rappert and Balmer 2007). Interfering with the openness of science was, they claimed, to strike at the very heart of what it meant to be scientific. Certainly, many reasons exist for rejecting 'sensitive but unclassified' and the thinking behind it (National Academy of Sciences 2004: 8), but in the context of this book it is interesting to note that the standard defence of science as an intrinsically open activity cannot be accepted at face value. Openness in science is not an equilibrium state of affairs: it needs to be argued for, put in place and actively maintained.

In addition to the specific debate over 'sensitive but unclassified', over the past decade civilian scientists have increasingly being called upon to self-censor their work and to foresee where it might be misused for malign purposes (National Academy of Sciences 2004, Rappert 2003, 2007a, Rappert and McLeish 2007, Royal Society-Wellcome 2004). A group representing major scientific publications and author-groups in the life sciences recently published a Statement of Scientific Publication and Security in the leading journals *Science, Nature* and *Proceedings of the National Academy of Sciences* (Journal Editors and Authors Group 2003). The statement noted that in extreme cases an editor could modify or even refuse to publish an article that could be used for malign purposes. Additionally, The US National Archives and Records Administration (NARA) announced that 'in light of the terrorist events of September 11, we are re-evaluating access to some previously open archival materials and reinforcing established practices on screening materials not yet open for research'.[2] In these instances, complex judgements have to be made about elements of socio-technical systems and whether or not to restrict access to information. It cannot simply be assumed, for example, that a published paper about sequencing the genome of a pathogenic virus contains a recipe for anyone to follow regardless of skill, training or access to and mastery of specialised equipment (Vogel 2008a). In parallel with the VX case study in chapter 7, any such assessments over the potential uses of codified research outputs equally involve someone having to decide what is dangerous in relation to the characteristics, motives, expertise and resources of putative criminals, be they sub-state or state-level actors. It involves judging what fragments of knowledge are already 'out there', and what those fragments could add up to now or at any time in the future. Such an assessment does one of two things, it either calls for knowledge – or more accurately, intelligence – to which most scientists have no

2 http://archives.gov/research_room/whats_new/notices/access_and_terrorism.html accessed 13/07/05.

access, or as in the VX case, it forces them to invoke a formless spectre in lieu of 'the enemy'.

Secrecy also played a role in the build-up to the second Gulf War and 2003 invasion of Iraq. Kathleen Vogel has cast a sociological eye over the since discredited claims made by the Bush administration that Iraq possessed mobile biological weapons factories (Vogel 2008b). She provides a careful analysis of the different assumptions adopted by the US intelligence analysts, in particular focussing on a key informant who was codenamed 'Curveball'. Curveball's information eventually resulted in a series of diagrams and sketches of the alleged mobile factories, which had been worked up from Curveball's original drawings into more stylised versions. These were used in Secretary of State Colin Powell's crucial address to the United Nations in the run up to the invasion. Vogel contrasts the apparent certainty implied by these diagrams with activity in other sections of the CIA, where critics seriously questioned the reliability of Curveball as an informant. She argues that what was said, the 'technical information' supplied by Curveball, and the credibility of who was saying it became separated. This separation meant that secret disagreements did not surface in the public depictions of mobile biological weapons production facilities, thus leading to different degrees of certainty being attributed to Curveball's knowledge claims in different locations. Secrecy, as I also have argued in this book, has a geography that is far from inconsequential. Turning from these contemporary manifestations of secrecy and their parallels with aspects of the historical case studies, I suggest a number of points that might re-surface as more enduring features of secrecy.

Secrecy changes science. A close reading of events in the history of chemical and biological warfare research and policy demonstrates how it is difficult to maintain the idea of secrecy as simply a veil drawn around an immutable activity. Secrecy is enacted or performed; it produces, alters, re-configures whatever comes within its ambit. Different questions get asked, different lines of research are pursued, normative judgements are altered and actions are taken, such as tailing the *Carella* rather than directly intervening, all shaped by secrecy. These processes are at their most evident in the previous chapter's case study of VX, where we saw secrecy play a role in actively constructing knowledge (such as what is or is not dangerous, what is or is not hidden) as well as ignorance (such as the idea that nothing of importance happened). At the same time, secrecy proliferates. New measures to enforce secrecy have to be invoked to maintain secrets, particularly when their protection is threatened. While this is not a particularly novel observation, it becomes more interesting when coupled with the insights from a co-production perspective introduced in chapter 3. As discussed, this new social ordering also helped establish new knowledge about the behaviour of plague bacteria as the accident became an experiment of opportunity.

What is a secret? It is easy to construe a secret as something bounded: a thing or possession, an object in a box. From the outset I have argued that, while social actors frequently employ this categorisation of secrecy, it is not the only way to understand secrets, which can also be regarded in more relational terms. As we saw in chapter 4, the act of classification creates a system of more or less valuable secrets, at the same time reinforcing the identity of those who are either allowed or denied access. So, possessing a secret defines objects, knowledge and people in relation to others. At its most simple this is merely everything that is 'in' on the secret and then the rest – although this immediately requires much further clarification. Indeed, openness and secrecy are best not treated as polarised opposites. It seems obvious to state that there are instead degrees of secrecy, but less obvious that much of the way we think and talk about secrecy belies this spectrum. Things are either open or secret. We know or do not know. A number of my case studies – particularly in chapters 6 and 7 – point to the partiality of secrecy, its foregrounding or backgrounding. This leads us to think of a concentric circle as a zoned metaphor for secrecy and openness. When a secret is revealed, there is always going to be something left concealed whether by intent or accident. When intentional, as we saw with the cover stories in chapters 3 and 4 or the public relations exercises in chapter 6, this can be in order to create an illusion of transparency.

There is another interesting way in which openness and secrecy elide, which returns to my discussion of revelation and disclosure in chapter 1. Counter to arguments that secrets are simply forgotten about until the moment of revelation, when all of the concern and anxiety around secrecy becomes manifest, I conjectured that this moment of revelation can simply result in a second order debate about whether the revealed secret ever really was a secret. The discussion of the *Carella* and the VX patent in particular should have substantiated this earlier claim. In both cases, the realisation that a secret might have been breached entailed discussions about what was already known and by whom. For instance, taking the suggestion that because the details of the synthesis of V-agents had been published in technical literature in seven different countries, this was sufficient for Porton's officials to claim that the secret was not a secret.

This observation leads us back to a central question about secrecy and science that I posed in my preface and in chapter 1: who knows what, where? And, crucially, what is at a stake in this distribution of knowledge and ignorance? Throughout the case studies in this book, this has been both an analytical question posed by me but it is also one which the historical actors in my accounts have answered either explicitly or implicitly. Attributions of who knows what often take on particular significance. So, frequently we have encountered a situation – once again exemplified in the VX case study – where a problem arises and it is assumed that someone, somewhere, knows the answer. In this instance, what was in the patent found in a public library in 1975? Secrecy, here, maintained the illusion that somewhere there already was an answer to this question waiting to be found, whereas the documentary record reveals no such easy reading of the patent, even by expert chemists, and instead there was an on-going negotiation over how to interpret both the contents and the implications of that disputed content. In a slightly different vein, chapter 2 demonstrated how the isolation of biological weapons scientists meant that a certain 'moral economy' could be sustained with arguments aimed at an inferred, rather than actual, public. Other attributions of who knows what, where, have reversed stereotypes such as the scientist, normally portrayed as laying claim to certainty, as a legitimate doubter in chapter 5, and in chapter 3, the public, often portrayed as ignorant, as near-omniscient in their ability to recognise a biological weapons trial based on scant evidence.

Secret science is, above all, and to return to my earliest assertions at the start of this book, not simply the same as open science but just done behind closed doors. This point has been illustrated time and again in my empirical case studies. Taken together, the case studies show how secrecy matters for science, and what might be at stake when we consider the implications of masking or unmasking science. It is not simply about hidden bits of information. It is not simply about two groups: those who know everything about something but hide it, and those who remain ignorant and in the open. Science is altered by secrecy, and people change once they have access to secrets and the means to enforcing those secrets. In zoned social worlds with degrees of openness and secrecy, different people make different claims to knowledge, while different resources for assessing the credibility of those claims and the legitimacy of the claims-makers are brought to bear. All these observations soon spill over into topics that have been only tangential to my discussion, not least how we might understand phenomena such as conspiracies, censorship, privacy, surveillance and vetting. Indeed, it should be apparent by now that Simmel was entirely correct to claim that there was something fundamental for understanding social life in the concept of secrecy. The implications of his observation for understanding the sociology of science should be clearer now that we have considered how secrets are enacted within, and help to maintain, the fractured and uneven landscape of knowledge production.

# References


Adams, V. 1986. Chemical and biological warfare, in *Ethics and European Security*, edited by B. Paskins. London: Croom Helm.

Agar, J. 2008. What happened in the sixties? *British Journal for History of Science*, 41(4), 567–600.

Agar, J. and Balmer, B. 1998. British scientists and the Cold War: The Defence Policy Research Committee and information networks, 1947–1963. *Historical Studies in the Physical Sciences*, 28(2), 209–52.

Anon. 1985. Stewart questions government of germ warfare experiments. *Stornoway Gazette*, 27 July, no page number.

Arnold, L. 2001. *Britain and the H-Bomb*. Basingstoke: Palgrave.

Axelrod, R. and Hamilton, W.D. 1981. The evolution of cooperation. *Science*, New Series, 211(4489), 1390–6.

Bacon, F. 2006 [1624]. *The New Atlantis* (Gloucester: Dodo Press).

Badash, L. 2003. From security blanket to security risk: scientists in the decade after Hiroshima. *History and Technology*, 19(3), 241–56.

Balmer, B. 1997. The drift of biological weapons policy in the UK, 1945–1965. *Journal of Strategic Studies*, 20(4), 115–45.

Balmer, B. 2001. *Britain and Biological Warfare: Expert Advice and Science Policy, 1935–1965*. Basingstoke: Palgrave.

Balmer, B. 2006. The British program, in *Deadly Cultures: Biological Weapons Since 1945*, edited by M. Wheelis, L, Rózsa, and M. Dando. Cambridge Mass.: Harvard University Press, 47–83.

Balmer, B. 2010. Keeping nothing secret: United Kingdom chemical warfare policy in the 1960s. *Journal of Strategic Studies*, 33(6), 871–93.

Barry, A. 2006. Technological zones. *European Journal of Social Theory*, 9(2), 239–253.

Bauer, M. and Gregory, J. 2007. From journalism to corporate communication in post-War Britain, in *Journalism, Science and Society: Science Communication Between News And Public Relations*, edited by M.W. Bauer and M. Bucchi. London: Routledge, 33–54.

Beck, U. 1992. *Risk Society: Towards a New Modernity*. London: Sage.

Bellman, B. 1981. The paradox of secrecy. *Human Studies*, 4(1), 1–24.

Berridge, V. 2003. Public or policy understanding of history? *Social History of Medicine*, 16(3), 511–23.

Biagioli, M. 2006. *Galileo's Instruments of Credit: Telescopes, Images, Secrecy*. Chicago: University of Chicago Press.

Blank, L. 2009. Two schools for secrecy: defining secrecy from the works of Max Weber, Georg Simmel, Edward Shils and Sissela Bok, in *Government*

*Secrecy: Classic and Contemporary Readings*, edited by S.L. Maret, S.L. and J. Goldman. Westport: Libraries Unlimited, 59–68.

Bok, S. 1989. *Secrets: On the Ethics of Concealment and Revelation*. New York: Vintage.

Bowker, G.C. 1994. What's in a patent?, in *Shaping Technology/Building Society: Studies in Sociotechnical Change*, edited by W. Bijker and J. Law. Cambridge Mass.: MIT Press, 53–74.

Bowker, G.C. and Starr, S.L. 1999. *Sorting Things Out: Classification and Its Consequences*. Cambridge Mass.: MIT Press.

Bratich, J. 2006. Public secrecy and immanent security: a strategic analysis. *Cultural Studies*, 20(4–5), 493–511.

Bucchi, M. 2004. *Science in Society: An Introduction to Social Studies of Science*. London: Routlege.

Buchanan, T. 2001. The courage of Galileo: Joseph Needham and the 'germ warfare' allegations in the Korean War. *History*, 86(284), 503–22.

Burhop, E.H.S. 1968. Foreword in *CBW, Chemical and Biological Warfare: Its Scope, Implications, and Future Development. The London Conference on CBW*, edited by S. Rose. London: Harrap, 5.

Burke, P. 2005. *History and Social Theory*. 2nd edition. Cambridge: Polity.

Campbell, B.L. 1985. Uncertainty as symbolic action in disputes among experts. *Social Studies of Science*, 15(3), 429–53.

Carmeli, Y.S. and Birenbaum-Carmeli, D. 2000. Ritualizing the 'natural family': secrecy in Israeli donor insemination. *Science as Culture*, 9(3), 301–24.

Carter, G.B. 2000. *Chemical and Biological Defence at Porton Down 1916–2000*. London: TSO.

Carter, G.B. and Pearson, G. 1996. North Atlantic chemical and biological research collaboration: 1916–1995. *Journal of Strategic Studies*, 19(1), 74–103.

Carter, G.B. and Pearson, G. 1999. British biological warfare and biological defence, 1925–1945, in *Biological and Toxin Weapons: Research, Development and Use from the Middle Ages to 1945*, edited by E. Geissler and J.E. van Courtland Moon. Oxford: Oxford University Press, 168–89.

Caute, D. 1988. *'68: The Year of the Barricades*. London: Paladin Books.

Chalk, R. 1985. Overview: AAAS Project on openness and secrecy in science and technology. *Science, Technology and Human Values*, 10(2), 28–35.

Charles, D. 2005. *Between Genius and Genocide: The Tragedy of Fritz Haber, Father of Chemical Warfare*. London: Jonathan Cape.

Chevrier, M. 2006. The Politics of Biological Disarmament, in *Deadly Cultures: Biological Weapons Since 1945*, edited by M. Wheelis, L, Rózsa, and M. Dando. Cambridge Mass.: Harvard University Press, 304–28.

Cloud, J. 2001. Imaging the world in a barrel: CORONA and the clandestine convergence of the earth sciences. *Social Studies of Science*, 31(2), 231–51.

Cloud, J. and Clarke, K. 1999. Through a shutter darkly: the tangled relationships between civilian, military and intelligence remote sensing in the early US space program, in *Secrecy and Knowledge Production*, edited by J. Reppy. Cornell University Peace Studies Program. Occasional Paper No.23, 35–56.

Cohn, C. 1987. Sex and death in the rational world of defense intellectuals. *Signs* 12(4), 687–718.

Cole, L. 1996. *The Eleventh Plague: The Politics of Biological and Chemical Warfare*. New York: WH Freeman and Co.

Cole, L. 2003. *The Anthrax Letters: A Medical Detective Story*. Washington DC: Joseph Henry Press.

Coleman, K. 2005. *A History of Chemical Warfare*. Basingstoke: Palgrave.

Collingridge, D. and Reeve, C. 1986. *Science Speaks to Power: The Role of Experts in Policymaking*. London: Pinter.

Collins, H. 1985. *Changing Order: Replication and Induction in Scientific Practice*. Chicago: Chicago University Press.

Collins, H. and Pinch, T. 1998. A clean kill? The role of Patriot in the Gulf War, in *The Golem at Large: What You Should Know About Technology*, by H. Collins and T. Pinch. Cambridge: Cambridge University Press, 7–29.

Colwell, R. and Zilinskas, R. 2000. Bioethics and the prevention of biological warfare, in *Biological Warfare: Modern Offence and Defence*, edited by R. Zilinskas. Boulder: Lynne Rienner, 225–45.

Crease, R.P. 2003. Fallout: issues in the study, treatment and reparations of exposed Marshall islanders, in *Science and Other Cultures: Issues in Philosophies of Science and Technology*, edited by R. Figueroa, R and S. Harding. London: Routledge, 106–25.

Croddy, E., Perez-Armendariz, C. and Hart, J. 2001. *Chemical and Biological Warfare: A Comprehensive Survey for the Concerned Citizen*. New York: Springer.

Crook, S., Pakulski, J. and Waters, M. 1992. *Postmodernization: Change in Advanced Society*. London: Sage.

Deleuze, G. and Guattari, F. 1987. *A Thousand Plateaus: Capitalism and Schizophrenia*. London: Continuum.

Dennis, M.A. 1994. 'Our first line of defense': two university laboratories in the postwar American state. *Isis*, 85(3), 427–55.

Dennis, M.A. 2007. Secrecy and science revisited: from politics to historical practice and back, in *The Historiography of Contemporary Science, Technology and Medicine: Writing Recent Science*, edited by R.E. Doel, and T. Söderqvist. London: Routledge, 172–84.

Department of the Army. 2005. *Classification of Former Chemical Warfare, Chemical and Biological Defense, and Nuclear, Biological and Chemical Contamination Survivability Information*, Army Regulation 380–86. Washington DC: Department of the Army.

Derrida, J. and Ferraris, M. 2001. *A Taste for the Secret*. Cambridge: Polity.

Donovan, C. 2005. The governance of social science and everyday epistemology. *Public Administration*, 83(3), 597–615.

Durodié, B. 2004. Facing the possibility of bioterrorism. *Current Opinion in Biotechnology*, 15(3), 264–8.

Eden, L. 2004. *Whole World on Fire: Organizations, Knowledge and Nuclear Weapons Devastation*. Ithaca: Cornell University Press.

Edgerton, D. 1990. Science and war, in *The Companion to the History of Modern Science*, edited by R.C. Olby et al. London: Routledge, 935–45.

Edgerton, D. 2006. *Warfare State: Britain 1920–1970*. Cambridge: Cambridge University Press.

Empson, R. 2007. Separating and containing people and things in Mongolia, in *Thinking Through Things: Theorising Artefacts Ethnographically*, edited by A. Henare, M. Holbraad, and S. Wastell. London: Routledge, 113–40.

Englehardt Jr, T. and Caplan, A. 1987. *Scientific Controversies: Case Studies in the Resolution and Closure of Disputes in Science and Technology*. Cambridge: Cambridge University Press.

Erikson, M. 2005. *Science, Culture and Society: Understanding Science in the 21st Century*. Cambridge: Polity.

Evans, G. and Durant, J. 1995. The relationship between knowledge and attitude in the public understanding of science in Britain. *Public Understanding of Science*, 4(1), 57–74.

Evans, R. 1997. Soothsaying or science? Falsification, uncertainty and social change in macroeconomic modelling. *Social Studies of Science*, 27(3), 395–438.

Evans, R. 2001. *Gassed: British Chemical Warfare Experiments on Humans at Porton Down*. London: House of Stratus.

Farrar-Hockley, A. 1995. *The British Part in the Korean War: Volume II, An Honourable Discharge*. London: HMSO.

Fox Keller, E. 1993. *Secrets of Life, Secrets of Death: Essays on Gender, Language and Science*. London: Routledge.

Galbraith, R. 2000. *Inside Outside: The Biography of Tam Dalyell – The Man They Can't Gag*. Edinburgh: Mainstream Publishing.

Galison, P. 2004. Removing knowledge. *Critical Inquiry*, 31(1), 229–43.

Geissler, E. ed. 1986. *Biological and Toxin Weapons Today*. Oxford: Oxford University Press.

Geissler, E. 1999. Biological warfare activities in Germany 1923–45, in *Biological and Toxin Weapons: Research, Development and Use from the Middle Ages to 1945*, edited by E. Geissler and J.E. van Courtland Moon. Oxford: Oxford University Press, 91–126.

Gentile, M. 2004. Former closed cities and urbanisation in the FSU: an exploration in Kazakhstan. *Europe-Asia Studies*, 56(2), 263–78.

Gibbs, D.N. 1995. Secrecy and International Relations. *Journal of Peace Research*, 32(2), 213–28.

Giddens, A. 1990. *The Consequences of Modernity*. Cambridge: Polity Press.

Gieryn, T.F. 1983. Boundary work and the demarcation of science from non-science: strains and interests in the professional ideologies of scientists. *American Sociological Review*, 48(6), 781–95.

Gieryn, T.F. 1995. Boundaries of science, in *Handbook of Science and Technology Studies*, edited by S. Jasanoff et al. London: Sage, 393–443.

Gieryn, T.F. 1999. *Cultural Boundaries of Science: Credibility on the Line*. Chicago: Chicago University Press.

Gieryn, T.F. 2000. A place for space in sociology. *Annual Review of Sociology*, 26, 463–96.

Gilbert, J. 2007. Public secrets: 'being with' in an era of perpetual disclosure. *Cultural Studies*, 21(1), 22–41.

Gilles, D. 2003. Probability and uncertainty in Keynes' *The General Theory*, in *The Philosophy of Keynes' Economics. Probability, Uncertainty and Convention*, edited by J. Runde and S. Mizuhara. London: Routledge, 111–29.

Glover, J. 1999. *Humanity: a Moral History of the Twentieth Century*. London: Jonathan Cape.

Golinski, J. 1988. The secret life of an alchemist, in *Let Newton Be!*, edited by J. Fauvel et al. Oxford: Oxford University Press, 147–167.

Goodman, J., McElligott, A. and Marks, L. eds. 2003. *Useful Bodies: Humans in the Service of Twentieth Century Medicine*. Baltimore: Johns Hopkins University Press.

Goodman, M. 2003. British intelligence and the Soviet atomic bomb, 1945–1950. *Journal of Strategic Studies*, 26(2), 120–151.

Goodman, M. 2004. Santa Klaus? Klaus Fuchs and the nuclear weapons programmes of Britain, the Soviet Union and America. *Prospero: The Journal of British Rocketry and Nuclear History*, 1(1), 47–70.

Goodman, M. 2005. Who is trying to keep what secret from whom and why? MI5-FBI relations and the Klaus Fuchs case. *Journal of Cold War Studies*, 7(3), 124–46.

Gould, C. and Hay, A. 2006. The South African biological weapons program, in *Deadly Cultures: Biological Weapons Since 1945*, edited by M. Wheelis, L, Rózsa, and M. Dando. Cambridge Mass.: Harvard University Press, 191–212.

Gowing, M. 1974. *Independence and Deterrence: Britain and Atomic Energy 1945–52*. Volume 1. London: MacMillan.

Gray, B.H. 1975. *Human Subjects in Medical Experimentation: A Sociological Study of the Conduct and Regulation of Clinical Research*. New York: Wiley-Interscience.

Greenberg, J. 2008. *From Betamax to Blockbuster: Video Stores and the Invention of Movies on Video*. Cambridge Mass.: MIT Press.

Gregory, J. 2000. The friendly argument: cosmological controversy as scientific sociability, paper presented at *Demarcation Socialised: or, Can We Recognise Science When We See It?* University of Cardiff, 24–27 August 2000.

Gregory, J. and Miller, S. 1998. *Science in Public: Communication, Culture and Credibility*. New York: Plenum.

Grint, K. and Woolgar, S. 1997. *The Machine at Work: Technology, Work and Organization*. Cambridge: Polity.

Grove, E.J. 1987. *Vanguard to Trident: British Naval Policy Since World War II*. London: Bodley Head.

Gross, M. 2007. The unknown in process: dynamic connections of ignorance, non-knowledge and related concepts. *Current Sociology*, 55(5), 742–59.

Guillemin, J. 2005. *Biological Weapons: From the Invention of State-Sponsored Programs to Contemporary Bioterrorism*. New York: Columbia University Press.

Guillemin, J. 2006. Scientists and the history of biological weapons: a brief historical overview of the development of biological weapons in the twentieth century. *EMBO Reports*, 7(SI), S45–S49.

Guillemin, J. 2008. Imperial Japan's germ warfare: the suppression of evidence at the Tokyo war crimes trial, 1946–48, in *Terrorism, War, or Disease? Unraveling the Use of Biological Weapons*, edited by A.L. Clunan et al. Stanford: Stanford University Press.

Guillemin, J. 2011. *American Anthrax: Fear, Crime and the Investigation of the Nation's Deadliest Bioterror Attack*. New York: Henry Holt.

Gusterson, H. 1996. *Nuclear Rites: A Weapons Laboratory at the End of the Cold War*. Berkeley: University of California Press.

Gusterson, H. 2003. The death of the authors of death: prestige and creativity among nuclear weapons scientists, in *Scientific Authorship: Credit and Intellectual Property in Science*, edited by M. Biagioli and P. Galison. New York: Routledge, 281–307.

Haber, L.F. 1986. *The Poisonous Cloud: Chemical Warfare in the First World War*. Oxford: Clarendon Press.

Haldane, J. 1987. Ethics and biological warfare. *Arms Control*, 8(1), 24–35.

Haldane, J.B.S. 1925. *Callinicus: A Defence of Chemical Warfare*. London: Kegan Paul, Trench, Trubner & Co.

Hammond, P.M and Carter, G.B. 2002. *From Biological Warfare to Healthcare: Porton Down, 1940–2000*. Basingstoke: Palgrave.

Harris, R. and Paxman, J. 1982. *A Higher Form of Killing: The Secret Story of Gas and Germ Warfare*. London: Chatto and Windus.

Harris, S. 1994. *Factories of Death: Japanese Biological Warfare 1932–45 and the American Cover-up*. London: Routledge.

Hart, J. 2006. The Soviet Biological Weapons Program, in *Deadly Cultures: Biological Weapons Since 1945*, edited by M. Wheelis, L. Rózsa, and M. Dando. Cambridge Mass.: Harvard University Press, 132–156.

Haynes, R.D. 1994. *From Faust to Strangelove: Representations of Scientists in Western Literature*. Baltimore: Johns Hopkins University Press.

Hennessy, P. 2001. *Whitehall*. London: Pimlico.

Hennessy, P. 2002. *The Secret State: Whitehall and the Cold War*. London: Allen Lane.

Hilgartner, S. 2000. *Science on Stage: Expert Advice as Public Drama*. Stanford: Stanford University Press.

Holden Reid, B. 1998. *Studies in British Military Thought: Debates With Fuller and Liddell Hart*. Nebraska: University of Nebraska Press.

Hughes, J. 2002. *The Manhattan Project: Big Science and the Atom Bomb*. Cambridge: Icon.

Irwin, A. 2001. *Sociology and the Environment*. Cambridge: Polity.

Jasanoff, S. 1987. Contested boundaries in policy-relevant science. *Social Studies of Science*, 17(2), 195–230.

Jasanoff, S. 1990. *The Fifth Branch: Science Advisers as Policymakers*. Cambridge Mass.: Harvard University Press.

Jasanoff, S. 2003. Ordering knowledge, ordering society, in *States of Knowledge: The Co-Production of Science and Social Order*, edited by S. Jasanoff. London: Routledge, 13–45.

Jefferson, C. 2009. *The Taboo of Chemical and Biological Weapons: Nature, Norms and International Law*, unpublished DPhil dissertation, SPRU, University of Sussex.

Jones. J.H. 1993. *Bad Blood: The Tuskegee Syphilis Experiment*. New York: Free Press.

Jones, S. 2010. *Death in a Small Package: A Short History of Anthrax*. Baltimore: The Johns Hopkins University Press.

Journal Editors and Authors Group. 2003. Statement of scientific publication and security. *Science*, 299(5610), 1149 (also published in *Nature* and *Proceedings of the National Academy of Sciences*).

Kaiser, D. 2005. The atomic secret in red hands? American suspicions of theoretical physicists during the early cold war. *Representations*, 90(1), 28–60.

Kenney, M. 1986. *Biotechnology: The University-Industrial Complex*. New Haven: Yale University Press.

Kenyon, I. 2000. Chemical weapons in the twentieth century: their use and their control. *The CBW Conventions Bulletin*, 48, 1–15.

Keynes, J.M. 1937. The general theory of employment. *Quarterly Journal of Economics*, 51(2), 209–23.

Knight, P. 2000. *Conspiracy Culture – from Kennedy to the X-Files*. London: Routledge.

Knorr-Cetina, K.D. 1981. *The Manufacture of Knowledge: An Essay on the Constructivist and Contextual Nature of Science*. Oxford and New York: Pergamon Press.

Knorr-Cetina, K.D. 1982. Scientific communities or transepistemic arenas of research? A critique of quasi-economic models of science. *Social Studies of Science*, 12(1), 101–30 .

Kohler, R. 1985. Bacterial physiology: the medical context. *Bulletin of The History of Medicine*, 59(1), 54–74.

Kohler, R. 1994. *Lords of the Fly: Drosophila Genetics and the Experimental Life*. Chicago: University of Chicago Press.

Kord, S. 2009. *Murderesses in German Writing, 1720–1860: Heroines of Horror*. Cambridge: Cambridge University Press.

Krohn, W. and Weyer, J. 1994. Society as a laboratory: the social risks of experimental research. *Science and Public Policy*, 21(3), 173–83.

La Follette, M. 1985. Secrecy in university-based research: who controls, who tells? *Science, Technology and Human Values*, Special Issue, 10(2), 3.

Lappé, M. 1990. Ethics in biological warfare research', in *Preventing A Biological Arms Race,* edited by S. Wright. Cambridge Mass.: MIT Press, 78–99.

Latour, B. 1983. Give me a laboratory and I will raise the world, in *Science Observed: Perspectives on the Social Studies of Science*, edited by K.D. Knorr-Cetina and M.J. Mulkay. London: Sage, 141–70.

Latour, B. 1987. *Science in Action*. Cambridge Mass.: Harvard University Press.

Latour, B. 1991. The impact of science studies on political philosophy. *Science, Technology and Human Values*, 16(1), 3–19.

Latour, B. and Woolgar, S. 1986. *Laboratory Life: The Construction of Scientific Facts*. Princeton: Princeton University Press.

Lederberg, J. ed. 1999. *Biological Weapons: Limiting the Threat*. Cambridge, Mass.: MIT Press.

Leigh, D. and Lashmar, P. 1985. British germ bomb sprayed trawler. *The Observer*, 21 July, 1.

Leitenberg, M. 1998. Resolution of the Korean War biological warfare allegations. *Critical Reviews in Microbiology*, 24(3), 169–94.

Levidow, L. 1990. Nuclear accidents by design. *Science as Culture*, 1(9), 99–109.

Livingstone, D. 2003. *Putting Science in Its Place: Geographies of Scientific Knowledge* .Chicago: University of Chicago Press.

Luhmann, N. 1994. Politicians, honesty and the higher amorality of politics. *Theory, Culture & Society*, 11(2), 25–36.

Lynch. M. 1998. The discursive production of uncertainty: The OJ Simpson "dream team" and the sociology of knowledge machine. *Social Studies of Science*, 28(5–6), 829–68.

Lynch, M. and Bogen, D. 1996. *The Spectacle of History: Speech, Text, and Memory at the Iran-Contra Hearings*. Durham: Duke University Press.

Lyotard, J. 1984. *The Postmodern Condition: A Report on Knowledge*. Minnesota: University of Minnesota Press.

MacDonald, C. 1990. *Britain and the Korean War*. Oxford: Blackwell.

MacKenzie, D. 1990. *Inventing Accuracy: A Historical Sociology of Nuclear Missile Guidance*. Cambridge Mass.: MIT Press.

MacKenzie, D. and Spinardi, G. 1995. Tacit knowledge, weapons design, and the uninvention of nuclear weapons. *American Journal of Sociology*, 101(1), 44–99.

MacLean, D. 1992. Ethics and biological defence research, in *The Microbiologist and Biological Defence Research: Ethics, Politics and International Security*, edited by R. Zilinskas. Annals of the New York Academy of Sciences, Vol.666. New York: New York Academy of Sciences, 100–12.

McElroy, R. 1991. The Geneva Protocol of 1925, in *The Politics of Arms Control Treaty Ratification*, edited by M. Krepon and D. Caldwell. New York: St Martin's Press, 125–66.

McGoey, L. 2007. On the will to ignorance in bureaucracy. *Economy and Society*, 36(2), 212–35.

McGoey, L. 2009. Pharmaceutical controversies and the performative value of uncertainty. *Science as Culture*, 18(2),151–64

McLeish, C. 1997. *The Governance Of Dual-Use Technologies In Chemical Warfare*. Unpublished MSc Dissertation, SPRU, University of Sussex.

McLeish, C. 2009. Opening up the secret city of Stepnogorsk: biological weapons in the former Soviet Union. *Area*,42(1), 60–9.

McLeish, C. and Balmer, B. 2012. Development of the V-series nerve agents, in *Innovation, Dual-Use and Security: Managing the Risks of Emerging Biological and Chemical Technologies*, edited by J.B. Tucker. Cambridge Mass.: MIT Press, chapter 20.

McLeish, C. and Nightingale, P. 2007. Biosecurity, bioterrorism and the governance of science: the increasing convergence of science and security policy. *Research Policy*, 36(10), 1635–54.

Marx, G. and Muschert, G. 2008. Simmel on secrecy: a legacy and inheritance for the sociology of information, in *Soziologie als Möglichkeit: 100 Jahre Georg Simmels Untersuchungen über die Formen der Vergesellschaftung [The Possibility of Sociology: 100 Years of Georg Simmel's Investigations into the Forms of Social Organisation]*, edited by C. Pailoud and C. Rol. Wiebsbaden Germany: VS Verlag für Sozialwissenschaften, 217–33.

Masco, J. 2001. Lie detectors: of secrets and hypersecurity in Los Alamos. *Public Culture*, 14(3), 441–67.

Mayor, A. 2003. *Greek Fire, Poison Arrows and Scorpion Bombs: Biological and Chemical Warfare in the Ancient World.* London: Duckworth.

Mendelsohn, E. 1997. Science, scientists and the military, in *Science in the Twentieth Century,* edited by J. Krige and D. Pestre. Amsterdam: Harwood,175–202.

Merton, R. 1973. The normative structure of science, in *The Sociology of Science: Theoretical and Empirical Investigations*, edited by N. Storer. Chicago: University of Chicago Press, 267–78.

Michael, M. 1994. Discourse and uncertainty: postmodern variations. *Theory and Psychology*, 4(3), 383–404.

Millstone, E. and van Zwanenberg, P. 2001. Politics of expert advice: lessons from the early history of the BSE saga. *Science and Public Policy*, 28(2), 99–112.

Mitchell, G. 2003. See an atomic blast and spread the word: indoctrination at ground zero, in *Useful Bodies: Humans in the Service of Twentieth Century Medicine*, edited by J. Goodman et al. Baltimore: Johns Hopkins University Press, 133–64.

Mol, A. 2002. *The Body Multiple: Ontology and Medical Practice*. Durham: Duke University Press.

Moon, JvC. 1992. Biological warfare allegations: the Korean case, in *The Microbiologist and Biological Defence Research: Ethics, Politics and International Security*, edited by R. Zilinskas. Annals of the New York Academy of Sciences, Vol. 666. New York: New York Academy of Sciences, 53–83.

Moreno, J.D. 2001. *Undue Risk: Secret State Experiments on Humans*. London: Routledge.

Morgan, M.S. 2003. Experiments without material intervention: model experiments, virtual experiments and virtually experiments, in *The Philosophy of Scientific Experimentation*, edited by H. Radder. Pittsburgh: University of Pittsburgh Press, 216–35.

Moynihan, D.P. 1999. *Secrecy: The American Experience*. New Haven: Yale University Press.

Mulkay, M. 1976. Norms and ideology in science. *Social Science Information*, 15(4–5), 637–56.

Mulkay, M. 1989. Don Quixote's double: a self-exemplifying text, in *Knowledge and Reflexivity: New Frontiers in the Sociology of Knowledge*, edited by S. Woolgar. London: Sage, 81–100.

Myers, G. 1995. From discovery to invention: the writing and re-writing of two patents. *Social Studies of Science*, 25(1), 57–105.

Nash, H.T. 1980. The bureaucratization of homicide, in *Protest and Survive*, edited by E. P. Thompson. Harmondsworth: Penguin, 62–74.

National Academy of Sciences. 2004. *Biotechnology Research in an Age of Terrorism: Confronting the Dual-Use Dilemma*. The 'Fink' Report. Washington DC: National Academies Press.

Naylor, S. 2005. Introduction: historical geographies of science – places, contexts, cartographies. *The British Journal for the History of Science*, 38(1), 1–12

Nelkin, D. ed. 1992. *Controversy: The Politics of Technical Decisions*. London: Sage.

O'Dell, T.H. 1994. *Inventions and Official Secrecy: A History of Patents in the United Kingdom*. Oxford: Clarendon Press.

Oakley, A. 2000 *Experiments in Knowing: Gender and Method in the Social Sciences*. Cambridge: Polity.

Paglen, T. 2009. *Blank Spots on the Map: The Dark Geography of the Pentagon's Secret World*. New York: Dutton.

Parish, J. and Parker, M. eds. 2001. *The Age of Anxiety: Conspiracy Theory and the Human Sciences* (Sociological Review Monographs). Oxford: Blackwell.

Park, K. 2006. *Secrets of Women: Gender, Generation, and the Origins of Human Dissection*. New York: Zone Books.

Parliamentary Office of Science and Technology (POST). 2009. Lessons from History, *Postnote* 323. Available at http://www.parliament.uk/documents/post/postpn323.pdf [accessed: 18 August 2011].

Perkins, C. and Dodge, M. 2009. Satellite imagery and the spectacle of secret spaces. *Geoforum*, 40(4),546–60.

Perry-Robinson, J. 1971. 'V-Agent Nerve Gases', in *The Problem of Chemical and Biological Warfare, Volume 1, The Rise of CB Weapons*, edited by J. Perry-Robinson and M. Leitenberg, SIPRI. Stockholm: Almqvist & Wiksell, 74.

Perry-Robinson, J. 1981. Chemical arms control and the assimilation of chemical weapons. *International Journal*, 36(3), 515–34.

Perry-Robinson, J. and Leitenberg, M. eds. 1971. *The Problem of Chemical and Biological Warfare, Volume 1, The Rise of CB Weapons*. SIPRI. Stockholm: Almqvist & Wiksell.

Piller, C. and Yamamoto, K.R. 1988. *Gene Wars: Military Control over the New Genetic Technology*. New York: Morrow.

Platt, J. 1988. What can case studies do? *Studies in Qualitative Methodology*, 1, 1–23.

Polanyi, M. 1958. *Personal Knowledge: Towards a Post-Critical Philosophy*. Chicago: Chicago University Press.

Potter, J. 1996. *Representing Reality: Discourse, Rhetoric and Social Construction*. London: Sage.

Powell, R. 2007. Geographies of science: histories, localities, practices, futures. *Progress in Human Geography*, 31(3), 309–29.

Price, R. 1995. A genealogy of the chemical weapons taboo. *International Organisation*, 49(1), 73–103.

Price, R. 1997. *The Chemical Weapons Taboo*. Cornell: Cornell University Press.

Proctor, R. 2006. 'Everyone knew but no one had proof': tobacco industry use of medical history expertise in US courts, 1990–2002. *Tobacco Control*, 15(supplement 4), 117–25.

Rappert, B. 2001. The distribution and the resolution of the ambiguities of technology; or why Bobby can't spray. *Social Studies of Science*, 31(4), 557–92.

Rappert, B. 2003. Coding ethical behaviour: the challenges of biological weapons. *Science and Engineering Ethics*, 9(4), 453–70.

Rappert, B. 2005. Prohibitions, weapons and controversy: managing the problems of ordering. *Social Studies of Science*, 35(2), 211–40.

Rappert, B. 2007a. *Biotechnology, Security and the Search for Limits: An Inquiry Into Research and Methods*. Basingstoke: Palgrave.

Rappert, B. 2007b. On the mid range: an exercise in disposing (or minding the gaps). *Science Technology & Human Values*, 32(6), 693–712.

Rappert, B. 2009. *Experimental Secrets: International Security, Codes, and the Future of Research*. Lanham: University Press of America.

Rappert, B. and Balmer, B. 2007. Rethinking 'secrecy' and 'disclosure': what science and technology studies can offer attempts to govern WMD threats', in *Technology and Security: Governing Threats in the New Millennium*, edited by B. Rappert. Basingstoke: Palgrave, 45–65.

Rappert, B. and McCleish, C. eds. 2007. *A Web of Prevention: Biological Weapons, Life Sciences and the Governance of Research*. London: Earthscan.

Ravetz, J.R. 1993. The sin of ignorance: ignorance of ignorance. *Knowledge: Creation, Diffusion, Utilization*, 15(2), 157–65.

Ravetz, J.R. and Funtowicz, S. 1993. Science for the Post-normal age. *Futures*, 25(7), 739–55.

Reardon, J. 2001. The Human Genome Diversity Project: a case study in coproduction. *Social Studies of Science*, 31(3): 357–88.

Reppy, J. ed. 1999. *Secrecy and Knowledge Production*. Cornell University Peace Studies Program.Occasional Paper No.23.

Resnik, D.B. 1998. *The Ethics of Science: An Introduction*. London: Routledge.

Rier, D.A. 1999. The versatile 'caveat' section of an epidemiology paper: managing public and private risk. *Science Communication*, 21(1), 3–37.

Roland, A. 2003. Science, technology, and war, in *The Cambridge History of Science. Volume 5: The Modern Physical and Mathematical Sciences*, edited by M. Nye. Cambridge: Cambridge University Press, 561–78.

Royal Society. 2000. *Measures for Controlling the Threat from Biological Weapons*. London: The Royal Society.

Royal Society. 2005. *The Roles of Codes of Conduct in Preventing the Misuse of Scientific Research*. London: The Royal Society.

Royal Society and Wellcome Trust. 2004. *Do No Harm: Reducing the Potential for the Misuse of Life Science Research*. Report of a Royal Society-Wellcome Trust meeting held at the Royal Society on 7 October 2004. London: The Royal Society.

Schmidt, U. 2004. *Justice at Nuremberg: Leo Alexander and the Nazi Doctors' Trial*. Basingstoke: Palgrave.

Schmidt, U. 2006. Cold War at Porton Down: informed consent in Britain's biological and chemical warfare experiments. *Cambridge Quarterly for Healthcare Ethics*, 15(4), 366–80.

Schmidt, U. 2007. Medical ethics and human experimentation at Porton Down: informed consent in Britain's biological and chemical warfare experiments', in *History and Theory of Human Experimentation. The Declaration of Helsinki and Modern Medical Ethics* edited by U. Schmidt and A. Frewer. Frankfurt and New York: Franz Steiner, 283–313.

Schmidt, U and Frewer, A. eds. 2007. *History and Theory of Human Experimentation: The Declaration of Helsinki and Modern Medical Ethics*. Frankfurt and New York: Franz Steiner.

Serber, R. 1998. *Peace and War: Reminiscences of a Life on the Frontiers of Science*. New York: Columbia University Press.

Shackley, S. and Wynne, B. 1996. Representing uncertainty in global climate change science and policy: boundary-ordering devices and authority. *Science, Technology and Human Values*, 21(3), 275–302.

Shapin, S. 1988. The house of experiment in seventeenth-century England. *Isis*, 79(3), 373–404.

Shapin, S. 1998. Placing the view from nowhere: historical and sociological problems in the location of science. *Transactions of the Institute of British Geographers*, 23(1), 5–12.

Shapin, S. and Schaffer, S. 1985. *Leviathan and the Air Pump: Hobbes, Boyle and the Experimental Life*. Princeton: Princeton University Press.

Shils, E. 1996 [1956]. *The Torment of Secrecy: Background and Consequences of American Security Policies*. Lanham: Ivan R Dee.

Shulinder, A. 1999. Learning to keep secrets: the military and a high-tech company, in *Secrecy and Knowledge Production*, edited by J. Reppy. Cornell University Peace Studies Program. Occasional Paper No.23, 77–92.

Simmel, G. 1906. The secret and the secret society. *American Journal of Sociology*, 11(4), 441–98.

Sims, N. 1987. Morality and biological warfare. *Arms Control*, 8(1), 5–23.

Sinsheimer, R.L. 1990. The responsibility of scientists, in *Preventing a Biological Arms Race*, edited by S. Wright. Cambridge Mass.: MIT Press, 71–7.

SIPRI. 1971. *The Problem of Chemical and Biological Warfare*. Stockholm: Almqvist & Wiksell. (6 Volumes).

Sismondo, S. 2010. *An Introduction to Science and Technology Studies*. 2nd Edition. Oxford: Blackwell.

Smithson, M. 1989. *Ignorance and Uncertainty: Emerging Paradigms*. New York: Springer-Verlag.

Somerville, M. and Atlas, R. 2005. Ethics: a weapon to counter bioterrorism. *Science*, 307(5717), 1881–2.

Spiers E.M. 2006. Gas disarmament in the 1920s: hopes confounded. *Journal of Strategic Studies*, 29(2), 281–300.

Star, S.L. 1985. Scientific work and uncertainty. *Social Studies of Science*, 15(3), 391–427.

Stirling, A. 2007. Science, precaution and risk assessment: towards more measured and constructive policy debate. *EMBO Reports*, 8, 309–15.

Stocking, H.S. 1998. On drawing attention to ignorance. *Science Communication*, 20(1), 165–78.

Stocking, H.S. and Holstein, L. 1993. Constructing and reconstructing scientific ignorance: ignorance claims in science and journalism. *Knowledge: Creation, Diffusion, Utilization*, 15(2), 186–210.

Stocking, H.S. and Holstein, L. 2009. Manufacturing doubt: journalists' roles and the construction of ignorance in a scientific controversy. *Public Understanding of Science*, 18(1), 23–42.

Strauss, H. and King, J. 1986. The fallacy of defensive biological weapons programmes, in *Biological and Toxin Weapons Today*, edited by E. Geissler. Oxford: Oxford University Press, 66–73.

Strydom, P. 2002. *Risk, Environment, Society: Ongoing Debates, Current Issues and Future Prospects*. Buckingham: Open University Press.

Szerszynski, B. 2005. Beating the unbound: political theatre in the laboratory without walls, in *Performing Nature: Explorations in Ecology and the Arts*, edited by G. Giannachi and N. Stewart. Frankfurt and New York: Peter Lang, 181–97.

Taussig, M. 1999. *Defacement: Public Secrecy and the Labor of the Negative*. Stanford: Stanford University Press.

Taylor, B. 1985. Isles plague tests shock. *Aberdeen Press and Journal*, 22 July, 2.

Toulmin, S. 1991. *Cosmopolis*. Chicago University Press.

Thorpe, C. 2002. Disciplining experts: scientific authority and liberal democracy in the Oppenheimer case. *Social Studies of Science*, 32(4), 525–62.

Thorpe, C. 2007. *Oppenheimer: The Tragic Intellect*. Chicago: University of Chicago Press.

Tucker, J. 2007. *War of Nerves: Chemical Warfare from World War I to Al-Qaeda*. New York: Anchor Books.

Tucker, J. and Mahan, E.R. 2009. *President Nixon's Decision to Renounce the US Offensive Biological Weapons Program*. Washington DC: National Defence University Press.

Turchetti, S. 2003. Atomic secrets and governmental lies: nuclear science, politics and security in the Pontecorvo case. *British Journal for the History of Science*, 36(4), 389–415.

Vaughan, D. 1997. *The Challenger Launch Decision: Risky Technology, Culture and Deviance at NASA*. Chicago: University of Chicago Press.

Vincent, D. 1998. *The Culture of Secrecy: Britain 1832–1998*. Oxford: Oxford University Press.

Vogel, K. 2006. Bioweapons proliferation: where science studies and public policy collide. *Social Studies of Science*, 36(5), 659–90.

Vogel, K. 2008a. Framing biosecurity: an alternative to the biotech revolution model? *Science and Public Policy*, 35(1), 45–54.

Vogel, K. 2008b. 'Iraqi Winnebagos™ of death': imagined and realized futures of US bioweapons threat assessments. *Science and Public Policy*, 35(8), 561–73.

Wang, J. 1998. *American Science in an Age of Anxiety: Scientists, Anticommunism, and the Cold War*. Carolina: University of North Carolina Press.

Weart, S.R. 1988. *Nuclear Fear: A History of Images*. Cambridge Mass.: Harvard University Press.

Weber, M. 2009 [1920]. Bureaucracy: characteristics and the power position of bureaucracy, in *Government Secrecy: Classic and Contemporary Readings*, edited by S.L. Maret and J. Goldman. Westport: Libraries Unlimited, 44–9.

Weinstein, M. 2001. A public culture for guinea pigs: US human research subjects after the Tuskegee study. *Science as Culture*, 10(2), 195–223.

Welsh, A. 1985. *George Eliot and Blackmail*. Cambridge Mass.: Harvard University Press.

Westwick, P.J. 2000. Secret science: a classified community in the national laboratories. *Minerva*, 38(4), 363–91.

Wheelis, M. 1999. Biological sabotage in World War I, in *Biological and Toxin Weapons: Research, Development and Use from the Middle Ages to 1945*, edited by E. Geissler, J.E. van Courtland Moon. Oxford: Oxford University Press, 35–62.

Wheelis, M., Rózsa, L. and Dando, M. eds. 2006. *Deadly Cultures: Biological Weapons Since 1945*. Cambridge Mass.: Harvard University Press.

White, L. 2000. Telling more: lies, secrets and history. *History and Theory*, 39(4), 11–22.

Williams, S.J. 2004. Bioattack or panic attack? Critical reflections on the ill-logic of bioterrorism and biowarfare in late/postmodernity. *Social Theory and Health*, 2(1), 67–93.

Willis, E.A. 2003. Seascape with monkeys and guinea-pigs: Britain's biological weapons research programme, 1948–1954. *Medicine, Conflict and Survival*, 19(4), 285–302.

Willis, E.A. 2004. Contamination and compensation: Gruinard as a 'menace to the mainland'. *Medicine, Conflict and Survival*, 20(4), 334–43.

Wilson, B. 1985. Comment. *West Highland Free Press*, 26 July, 3.

Woods, A. 2004. *A Manufactured Plague: The History of Foot-and-Mouth Disease in Britain*. London: Earthscan.

Wright, S. ed. 1990. *Preventing a Biological Arms Race*. Cambridge Mass.: MIT Press.

Wright, S. 2002. Geopolitical origins, in *Biological Warfare and Disarmament: New Problems/New Perspectives*, edited by S. Wright. Lanham: Rowman and Littlefield, 313–42.

Wright, S. and Wallace, D.A. 2002. Secrecy in the biotechnology industry: implications for the biological weapons convention, in *Biological Warfare and Disarmament: New Problems/New Perspectives*, edited by S. Wright. Lanham: Rowman and Littlefield, 369–90.

Wynne, B. 1987. Uncertainty – technical and social, in *Science for Public Policy*, edited by H. Brooks and C. Cooper. Oxford: Pergamon Press, 95–115.

Wynne, B. 1995. Public understanding of science, in *Handbook of Science and Technology Studies*, edited by S. Jasanoff et al. London: Sage, 361–88.

Yearley, S. 2005. *Making Sense of Science: Understanding the Social Study of Science*. London: Sage.

Young, J. 1996. *Winston Churchill's Last Campaign*. Oxford: Clarendon Press.

Zehr, S.C. 2000. Public representations of scientific uncertainty about global climate change. *Public Understanding of Science*, 9(2), 85–103.

Zilinskas, R.A. ed. 2000. *Biological Warfare: Modern Offence and Defence*. Boulder Colorado: Lynne Reiner Publishers.

van Zwanenberg, P. and Millstone, E. 2005. *BSE: Risk, Science and Governance*. Oxford: Oxford University Press.

# Index